Ministry of Defence Roles and Required Technical Resources

John F. Schank

Cynthia R. Cook

Robert Murphy

James Chiesa

Hans Pung

John Birkler

T0312714

Prepared for the
United Kingdom's Ministry of Defence

RAND EUROPE

The research described in this report was prepared for the United Kingdom's Ministry of Defence. The research was conducted jointly in RAND Europe and the RAND National Security Research Division.

Library of Congress Cataloging-in-Publication Data

The United Kingdom's nuclear submarine industrial base.
   p. cm.
   "MG-326/1."
   "MG-326/2."
   "MG-326/3."
   Includes bibliographical references.
   ISBN 0-8330-3797-8 (v. 1 : pbk.)—ISBN 0-8330-3845-1 (v. 2 : pbk.)—
ISBN 0-8330-3784-6 (v. 3 : pbk.)
   1. Nuclear submarines—Great Britain—Design and construction. 2.
Shipbuilding industry—Great Britain. 3. Military-industrial complex—Great
Britain. 4. Defense industries—Great Britain. I. Schank, John F. (John Frederic),
1946– II. Raman, Raj.

V859.G7.U55 2005
359.9'3834'0941—dc22

                                                                    2005010735

The RAND Corporation is a nonprofit research organization providing objective analysis and effective solutions that address the challenges facing the public and private sectors around the world. RAND's publications do not necessarily reflect the opinions of its research clients and sponsors.

RAND® is a registered trademark.

Cover design by Peter Soriano
Cover photo by Stephen Bloodsworth

© Copyright 2005 RAND Corporation

All rights reserved. No part of this book may be reproduced in any form by any electronic or mechanical means (including photocopying, recording, or information storage and retrieval) without permission in writing from RAND.

Published 2005 by the RAND Corporation
1776 Main Street, P.O. Box 2138, Santa Monica, CA 90407-2138
1200 South Hayes Street, Arlington, VA 22202-5050
201 North Craig Street, Suite 202, Pittsburgh, PA 15213-1516
RAND URL: http://www.rand.org/
To order RAND documents or to obtain additional information, contact
Distribution Services: Telephone: (310) 451-7002;
Fax: (310) 451-6915; Email: order@rand.org

Preface

Several recent trends have warranted concern about the future vitality of the United Kingdom's submarine industrial base. Force structure reductions and budget constraints have led to long intervals between design efforts for new classes and low production rates. Demands for new submarines have not considered industrial base efficiencies, resulting in periods of feast or famine for the organisations that support submarine construction. Government policies have resulted in a reduction in the submarine design and management resources within the Ministry of Defence (MOD) in an effort to reduce costs. Yet the aforementioned production inefficiencies coupled with increased nuclear oversight have resulted in greater costs.

Concerned about the future health of the submarine industrial base, the MOD asked RAND Europe to examine the following four issues:

- What actions should be taken to maintain nuclear submarine design capabilities?
- How should nuclear submarine production be scheduled for efficient use of the industrial base?
- What MOD capabilities are required to effectively manage and support nuclear submarine programmes?
- Where should nuclear fuelling occur to minimise cost and schedule risks?

This report addresses the third of those issues. The following companion reports address the other three:

- *The United Kingdom's Nuclear Submarine Industrial Base, Volume 1: Sustaining Design and Production Resources*, MG-326/1-MOD
- *The United Kingdom's Nuclear Submarine Industrial Base, Volume 3: Options for Initial Fuelling*, MG-326/3-MOD.

This report should be of interest not only to the Defence Procurement Agency and to other parts of the Ministry of Defence, but also to service and defence agency managers and policymakers involved in weapon system acquisition on both sides of the Atlantic. It should also be of interest to shipbuilding industry executives within the United Kingdom. This research was undertaken for the MOD's Attack Submarines Integrated Project Team jointly by RAND Europe and the International Security and Policy Center of the RAND National Security Research Division, which conducts research for the US Department of Defense, allied foreign governments, the intelligence community, and foundations.

For more information on RAND Europe, contact the president, Martin van der Mandele. He can be reached by email at mandele@rand.org; by phone at +31 71 524 5151; or by mail at RAND Europe, Newtonweg 1, 2333 CP Leiden, The Netherlands. For more information on the International Security and Defense Policy Center, contact the director, Jim Dobbins. He can be reached by email at James_Dobbins@rand.org; by phone at (703) 413-1100, extension 5134; or by mail at RAND Corporation, 1200 South Hayes Street, Arlington, VA 22202-5050. More information about RAND is available at www.rand.org.

Contents

Figures and Tables

Figures

Tables

Summary

The United Kingdom's Ministry of Defence (MOD) is currently procuring a new class of nuclear-powered attack submarines, the Astute. For various reasons, the Astute programme has fallen behind schedule and has exceeded early cost estimates. Some of these divergences may stem from the acquisition strategy used for this class, which represented a break from how submarines were historically acquired. Here we determine what MOD institutional resources would be required and how they should be organised and brought to bear to maximise product quality and minimise cost and schedule penalties in submarine acquisition. We base our analyses on extensive interviews with MOD and contractor personnel concerned with submarine acquisition, along with the literature on partnering and on best practices in acquisition.

Overview of Submarine Acquisition To Date

For all classes of nuclear submarines, from Valiant up through Vanguard, the MOD exercised significant authority and responsibility in design and development and performed the integration role for the acquisition programme. It completed the concept definition, managed the design process, and maintained design authority[1] through-

---

[1] Joint Service Publication 430 defines *design authority* as 'An organisation with the professional competence and authority to specify design requirements, undertake design tasks, apply configuration management to designs and associated documentation, whilst con-

out the life of the submarine. This involved considerable, ongoing investment in programme management and technical skills, which were maintained with the regular introduction of new ship classes and subsequent maintenance and upgrade of in-service submarines.

The MOD by no means performed all the necessary functions required to design a complete submarine in-house. For example, it hired a shipbuilder, Vickers Shipbuilding and Engineering Limited of Barrow-in-Furness, to perform the detailed design work and build the submarine. Various subcontractors designed major systems and subsystems.

This acquisition structure lasted for three decades, as the United Kingdom established and worked towards a nuclear submarine building programme with a regular production schedule. Two major changes during the 1990s influenced the way the MOD managed its nuclear submarine acquisition programmes. First, the MOD felt pressure to reduce its staffing levels and transfer major programme responsibilities to the prime contractor. This pressure stemmed from the change in national political philosophy, towards a smaller role for government in society, that had been in the process of implementation for a number of years, as well as from criticisms that the administrative infrastructure used to manage defence system acquisition was too large and costly. Whole organisations that played a major role in submarine acquisition were eliminated.

Second, in the latter part of the decade, a new system for defence systems acquisition—now known as Smart Acquisition—was adopted. This system emphasised partnership with industry in requirements generation, the creation of integrated project teams (IPTs) to manage major defence system programmes, and the rotation of personnel to different programmes to broaden their exposure and knowledge.

As a result of these changes, the MOD followed a dramatically different model in acquiring the Astute class. Rather than being

---

tinuously monitoring the effectiveness of those activities for a given material state' (Ministry of Defence, *MOD Ship Safety Management: Issue 3, Part 1: Policy*, Ship Safety Management Office, July 2004b).

deeply involved in every stage of acquisition, the MOD sought to outsource as much of the work as possible to a prime contractor. The goal was to keep costs low[2] and to share the risk for any overrun with the prime contractor up to a point, after which the prime contractor would absorb all the cost risk. Also, the MOD gave the contractor more control over costs by conferring on it the authority to make design decisions. The transfer of responsibility to the contractor not only reflected concern about cost growth but also coincided with the loss of technical expertise at the MOD. Unfortunately, that loss, combined with the desire to avoid compromising the risk, resulted in the MOD not fully engaging with the contractor in important design decisions.

Although not the only cause, the new management model for Astute contributed to major cost and schedule problems. Currently, the programme is more than three years behind the schedule set during its initial stages and is approximately £1 billion over the initially approved cost.

The loss of internal submarine expertise, coupled with decreased nuclear submarine procurement rates, raises fundamental issues with respect to MOD capabilities as the ministry seeks to effectively oversee submarine design and production. The initial approach to Astute-class acquisition was an attempt to cope with a new philosophy and budgetary regime in which the MOD's capabilities had been reduced so much that its ability to oversee design and production was significantly affected. Recognising the cost and schedule problems confronting the Astute programme, the MOD proactively enhanced both resource levels and the degree of interaction and oversight in its relationship with the prime contractor. These actions have been effective to some degree; however, it is not clear whether they will be sufficient to handle an increased volume of activity that will occur during the design and build acceptance process in the next few years.

---

[2] Outsourcing activities does not, of course, eliminate the activity's cost to the MOD; the theory is that outsourcing lowers costs by increasing the efficiency with which the outsourced activities are accomplished.

Research Objectives and Approach

Given the various advantages and disadvantages of the greatly different models used by the MOD in managing nuclear submarine design and construction over the past few decades, the Attack Submarines Integrated Project Team (ASM-IPT) asked the RAND Corporation to address the following questions:

- What are the appropriate roles and functions for the MOD if it is to be a smart buyer of nuclear submarines?
- What management structure and level of additional resources are needed to perform those roles and functions?
- How could the MOD transition its workforce to the desired end goal?

To address these questions, RAND examined the historical record of UK submarine procurement to look for lessons from experience with the two approaches used to date—the model used through the Vanguard class and the initial model used for the Astute class. We conducted a wide range of interviews with various people in different MOD organisations involved with submarine design and acquisition as well as with private organisations such as BAE Systems Submarine Division, the current owner of the Barrow shipyard. Information and insights were also gathered from similar organisations in the United States.

Towards a New Acquisition Management Model

From the point of view of the economics, economic sociology, and management literatures, the evolution of submarine acquisition might be summarised as follows. Through the Vanguard class, the MOD's approach required large technical organisations and hence high costs in maintaining this internal bureaucracy. In the Astute era of increased outsourcing, a problematic working environment arose from the difficulty of writing a completely specified contract for a product as complex as a nuclear submarine. The MOD's attempt to seat the management of risk at the contractor rather than manage it in-house has also proved impossible to execute. The MOD is ultimately responsible for obtaining specified military performance, maintaining safety of operations, and delivering on schedule and at cost. Holding the contractor at risk for cost might mean driving it out of business, which would not accomplish the MOD's goal of providing a safe, militarily capable submarine to the flotilla.

To manage these risks, we propose a middle-ground alternative approach between the two acquisition models used in the past—that is, one model marked by a large internal bureaucracy and the other marked by an attempt to outsource key responsibilities. The proposed 'partnership' approach is broadly supported by the management best-practices literature on the benefits of intermediate structures, including joint ventures. While such an approach is one of the stated goals of Smart Acquisition, structures and processes to support a true partnership were not emplaced during the early years of the Astute programme.

The best-practices literature on purchasing and supplier management also suggests such a partnership model for acquisitions that are both high value and high risk—and nuclear submarines meet both of these criteria. Reported benefits from partnering include reduced cost, improved quality, and increased innovation. We describe what such a partnership model means, including the details of enhanced MOD engagement through all the submarine acquisition phases, and we suggest how current capabilities can be improved to support this new model.

Summary of Recommendations

Our recommendations are principally concerned with improving the MOD's ability to manage its risks and responsibilities in submarine acquisition. They are summarised as follows:

- Increase integration amongst MOD components, and increase partnering with industry in submarine design and development.

- Resolve critical safety and technical issues early in the programme by requiring early involvement of the Naval Authorities.
- Increase the ASM-IPT's engagement in design and construction at Barrow to improve the MOD's understanding of the contractor's detailed design and build performance and to facilitate the ministry's active participation in the acceptance process.
- Ensure through-life ship safety, maintenance and postdelivery control of design intent, and the propagation of lessons learned across submarine classes by transitioning design authority back from the prime contractor when the submarine enters service.
- Improve the MOD's management of technical support for nuclear propulsion throughout the fleet by shifting procurement and oversight of future nuclear steam-raising plant (NSRP) components and design services from being contractor furnished to being government furnished.

Note that these principles do not envision a return to a Vanguard-type model of MOD involvement. Design authority can remain with the contractor until the ship enters service. The contractor would still carry out most of the design effort.

These principles should be supported by a moderate growth of approximately 20 to 40 trained and experienced designers and engineers spread throughout the MOD's submarine-related organisations and by changes in the MOD's career management processes to encourage the development of submarine specialists.

In the remainder of this summary, we suggest appropriate roles for the MOD within its partnership with the prime contractor for each of the phases of submarine acquisition. We set out an organisational scheme, consistent with the current plan, by which the MOD might effectively meet its responsibilities under the partnership model. We then turn to other changes that would benefit the MOD, including modest staff increases and revision of the current career management philosophy. Finally, we present some thoughts on the transition to these desired end states.

MOD Roles Through the Phases of Submarine Acquisition

If the MOD is to meet its responsibilities for delivering a militarily capable, safe submarine to the flotilla on schedule and at a reasonable cost, it must fill certain roles during each phase of submarine acquisition, as seen in Table S.1. These roles involve various levels of leadership and partnership with the prime contractor.

The MOD is fully responsible for developing top-level military requirements, as laid out in the User Requirements Document (URD). The Director of Equipment Capability (DEC), also known as Customer 1, currently and appropriately performs this role. Approaches to meet these requirements are then worked out in concept and feasibility studies. The Future Business Group (FBG), working with the DEC, the Royal Navy flotilla (Customer 2), the Nuclear Propulsion IPT (NP-IPT), the Submarine Support IPT (SUB-IPT), and the Naval Authorities, should manage the initial phases of these studies. The FBG should also be supported by study contracts with relevant consulting firms and the shipbuilder(s). Once a new IPT is formed (or an existing IPT is tasked) to oversee the new programme, they should manage specification development, again with input from key stakeholders in the MOD, including the DEC, Customer 2, the SUB-IPT, the NP-IPT, the Naval Authorities, and

Table S.1
Recommended MOD Responsibilities
by Acquisition Phase

Acquisition Role	MOD Role
Requirements generation	Lead
Initial studies	Lead partner
Detailed design	Follow partner
Construction	Follow partner
Acceptance	Lead partner
Support	Lead

industry. This results in the System Requirements Document (SRD). The new IPT should work closely with the shipbuilder and other key suppliers to ensure mutual understanding of the SRD.

During detailed design and construction, the shipbuilder takes the lead. The MOD does not currently have the capability to design or build a submarine, roles that it clearly prefers to leave with industry. The shipbuilder must also lead vendor involvement where needed. However, the MOD should adopt the goal of working closely with—engaging with—the shipbuilder to address design issues and resolve problems as early in the process as possible to minimise cost of any necessary design or construction changes.[3]

One key to this approach is the development of formal and agreed-on procedures that include early communication with the Naval Authorities and other steps to manage critical, safety-related risks up front, rather than relying on the predelivery certification processes. These procedures should permit keeping the design authority and the majority of the effort at the shipbuilder.[4] All parties should communicate freely with MOD regulatory bodies.

Finally, when the submarine enters service and the MOD assumes the operational role, design authority should transfer back to the government. The MOD can choose to contract design authority support to private industry, acting as a design agent with formally prescribed roles. Every effort should be made to ensure that, for effectiveness, lessons learned from any class are transferred through the entire submarine fleet and that, for efficiency, fleetwide management practices are used.

---

[3] The MOD also needs to advise regarding government-furnished equipment, in-service experience with systems, and planned uses of the submarine.

[4] Some may question whether it is appropriate to let the contractor hold design authority, which lets it make important decisions that could be inappropriately swayed by commercial interests. Engagement between the ASM-IPT representative at Barrow and the contractor's design authority could provide assurance to the MOD that the design authority is exercised independently. Instead of bringing design authority back in-house, the MOD could invest in ensuring that the shipbuilder's processes are functioning correctly.

Revised MOD Organisational Structure

How might the MOD's organisational structure accommodate the roles discussed above? We agree with the recent establishment of a new nuclear organisation (similar in intent to the former Director General Submarines), which brings under one umbrella the four nuclear-related IPTs: the ASM-IPT, the NP-IPT, the Nuclear Weapons IPT (NW-IPT), and the SUB-IPT. In Figure S.1, we show this new organisation with the specific roles of the various groups involved in submarine programmes.

Figure S.1
Proposed MOD Organisation, Functions, and Interfaces

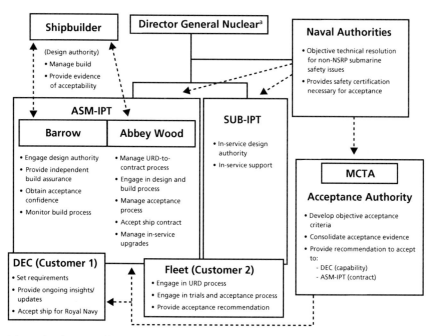

[a] Organizations not shown: NP-IPT and NW-IPT.
RAND *MG326/2-S.1*

Other Changes

Nuclear Steam-Raising Plant as Government-Furnished Equipment

The Astute contract marked the first time procurement and oversight of the NSRP were given to the prime contractor versus retaining them in the MOD. This change would yield benefits if the prime contractor could add value by managing the subcontract for provision of components and information necessary for nuclear regulatory requirements better than MOD had done previously when the NSRP was government-furnished equipment. However, we collected no evidence to show whether the prime contractor had managed Rolls-Royce better as a subcontractor than the MOD had done in the past. Even if BAE Systems were able to get priority in allocation of Rolls-Royce's limited technical resources in support of the Astute programme, this might not be the best allocation of Rolls-Royce resources from an overall MOD standpoint. The MOD also must manage responsibilities and risks with its operating fleet, refuellings, and refits at Devonport Management Limited along with NSRP support in Faslane. Having Rolls-Royce as a subcontractor to one or more MOD prime contractors, in addition to being an MOD prime contractor, invites suboptimisation of Rolls-Royce scarce technical resources and a diminution of the MOD's effectiveness and efficiency at managing its own risks. It appears that it would be a better approach to have the MOD actively involved in the Rolls-Royce technical resource allocation process, considering the MOD's overall nuclear propulsion management responsibilities. To that end, the MOD should shift NSRP components and design services to being government furnished on the Astute.

Increase MOD Staffing

The IPT's presence at Barrow is particularly important for improving MOD-contractor relations and to ensure product quality. Unfortunately, ASM-IPT staffing at Barrow may be insufficient for the impending high volume of work pertaining to the acceptance process in the next few years and should be augmented by possibly five personnel above the levels currently planned. We also recognise that staffing resources are limited at Abbey Wood, so if it proves impossible to increase the total staff, we recommend the transfer of Abbey Wood personnel to Barrow to the extent sufficient to support the Astute programme.

The staffing of the SUB-IPT, the NP-IPT, and the Naval Authorities should each be increased by approximately five to ten people with submarine expertise to support the increased interactions amongst the various organisations during the design and construction of a new class of submarines.

In summary, the roles and structures outlined above should be supported by an increase of approximately 20 to 40 trained and experienced personnel and by processes to support effective partnering and internal MOD linkages. Thus, the new management approach proposed here also depends on changes in career management and on process improvement and trust building.

Career Management

The MOD needs to aggressively manage its in-house technical expertise as it relates to acquisition and life-cycle management. The MOD's core technical expertise has been eroded through reduction in the ministry's workforce. It has also been depleted through a dual focus on job rotation for skill breadth and on the valuation of generalists, who can manage a contract, over specialists, who have more technical insight into design and construction. Mid-career and senior-level experts are still available, but the formal technical career track was eliminated some time ago. There are some indications that this lack has been recognised, and attempts are being made to remedy it. However, we are not convinced that the MOD has accepted the necessity of investing in the core of technical experts required to manage the responsibilities and risks inherent in submarine acquisition.

To resolve this, we recommend that the MOD reinstate an engineering career track for submarine-related skills. Career management might include further schooling and rotations at the shipyard in Barrow, within the different nuclear-related IPTs, with the Defence Procurement Agency's (DPA's) Sea Technology Group, and with the

Defence Logistics Organisation's submarine equipment IPTs. Currently, DPA staff are more likely to be promoted if they move away from technical roles and into management positions. A separate technical career track would provide room for growth for experts who choose to work as engineers throughout their careers.

Focus on Process Improvement and Building Trust

The relationship between the MOD and the current contractor has not been without its difficulties; each has raised concerns about the other's behaviour. The relationship has been characterised as one that is improving, but continuing problems suggest some residual bitterness and distrust.

While this is not something that can be cured immediately, a focus on the processes by which the shipbuilder and the government engage each other can lead to a climate of increased mutual understanding and eventually to trust. We suggest a task-based approach in which the MOD and the contractor work together to identify and resolve problematic areas.

Making the Transition

It is evident that the MOD faces continual pressure to cut staffing and reduce costs. Our recommendations do not require a significant increase in staffing, but some additions are suggested.

DPA management would need to be involved in any additional staffing and career structure changes. This represents a long-term staff planning issue and would likely require several years before it could effectively be implemented.

Process improvements can be managed from within the new nuclear organisation or at the ASM-IPT, with the engagement of the shipbuilder. This would require a strategic focus above and beyond the day-to-day issues that arise. Improving processes and strengthening the MOD–prime contractor relationship will be key to enhancing programme effectiveness without requiring a large increase in MOD staff.

Change is difficult because it can be threatening: It requires organisations and their members to accept that improvements over the current way of doing things are possible. In this study, we take only the first steps towards change—developing the case for it and indicating a vision for the future. The MOD now needs to further prepare the way forward and, in doing so, to remain conscious of how the organisation's culture, structure, and norms will react.

Acknowledgements

This research could not have been accomplished without the assistance of many individuals. Muir Macdonald, leader of the Attack Submarines Integrated Project Team, supported and encouraged the work. Numerous individuals in the Attack Submarines IPT offered information, advice, and assistance. Nick Hunt, Stephen Ranyard, Mark Hyde, Alisdair Stirling, Scot Butcher, and Commander Dai Falconbridge were especially helpful, providing background on the Astute programme and offering constructive criticism of interim findings and documentation. If we were to single out one individual who supported us in extraordinary ways, it would be Helen Wheatley, who provided data and information and facilitated our interactions with multiple organisations.

Commander Nigel Scott from the Director of Equipment Capability, Underwater Effects, offered advice and guidance on nuclear submarine requirements. Peter Duppa-Miller, Director of the Submarine Library, shared his extensive historical knowledge of UK submarine programmes and provided many documents that helped in our research. Commodore Mike Bowker, leader of the Nuclear Propulsion IPT, offered insights on the management of the nuclear aspects of the Astute contract. Tony Friday of the Future Business Group discussed his organisation's role in the development of submarine requirements.

Many individuals at BAE Systems Submarine Division shared their time and expertise with us and provided much of the data that were necessary to perform the analyses. Notable were Murray Easton,

Managing Director at Barrow; Huw James; Duncan Scott; John Hudson; and Tony Burbridge. We are deeply thankful for their assistance.

Numerous people from the US Navy and Electric Boat provided insights into the manner in which the United States addresses various aspects of submarine acquisition. Notable were RADM John Butler, Program Executive Officer for Submarines; Carl Oosterman, from the Nuclear Propulsion Directorate of the Naval Sea Systems Command; Captain Jeff Reed, of the Supervisor of Shipbuilding at Groton; and RADM Paul Sullivan. Millard Firebaugh, Fred Harris, Steve Ruzzo, Mark Gagnon, and Larry Runkle of Electric Boat shared their time and expertise.

Tony Bower of RAND and James Hall of the MOD provided helpful and constructive comments on an earlier version of the report. Phillip Wirtz did his usual exemplary job editing the document.

Of course, we alone are responsible for any errors contained in the report.

Abbreviations

ASM-IPT	Attack Submarines Integrated Project Team
BES	Babcock Engineering Services
CAD/CAM	computer-aided design/computer-aided manufacturing
CADMID	concept, assessment, demonstration, manufacture, in-service, and disposal
CSA	customer supplier agreement
CSMA	Captain Submarine Acceptance
CWG	Capability Working Group
DEC	Director of Equipment Capability
DGSM	Director General Submarines
DLO	Defence Logistics Organisation
DML	Devonport Management Limited
DOR(Sea)	Directorate of Operational Requirements (Sea)
DPA	Defence Procurement Agency
ECC	equipment capability customer
EVM	earned value management
FASM	Future Attack Submarine
FBG	Future Business Group
IPT	integrated project team
ITEA	integrated test, evaluation, and acceptance

ITT	invitation to tender
JSP	Joint Service Publication
MCTA	Maritime Commissioning Trials and Assessment
MOD	Ministry of Defence
MUFC	Maritime Underwater Future Capability
NATO	North Atlantic Treaty Organization
NP-IPT	Nuclear Propulsion Integrated Project Team
NSRP	nuclear steam-raising plant
NW-IPT	Nuclear Weapons Integrated Project Team
PFG	Pricing and Forecasting Group
PNO	Principal Naval Overseer
PWR2	Pressurised Water Reactor 2
S&T	science and technology
SRD	System Requirements Document
STG	Sea Technology Group
SUB-IPT	Submarine Support Integrated Project Team
TCE	transaction cost economics
URD	User Requirements Document
VSEL	Vickers Shipbuilding and Engineering Limited
WSA	Warship Support Agency
Y-ARD	Yarrow–Admiralty Research Department

CHAPTER ONE

Introduction

Background

The United Kingdom has been building nuclear submarines for well over 40 years. During that time, the approach to managing submarine acquisition programmes has shifted dramatically, from heavy involvement in every phase, to an arm's-length approach, and finally to one involving engagement with the prime contractor's design and shipbuilding personnel.

For the Valiant up through the Vanguard classes of nuclear submarines, the Ministry of Defence (MOD) had significant authority and responsibility in submarine design and development and performed an integration role for the acquisition programme. The MOD completed the initial concept design and maintained design authority[1] throughout the life of the submarine. This involved considerable, ongoing investment in programme management and in technical skills, which were maintained by the relatively frequent introduction of new submarine classes.

The MOD by no means performed all the necessary functions required to design a complete nuclear submarine in-house. It main-

---

[1] Joint Service Publication 430 defines *design authority* as 'An organisation with the professional competence and authority to specify design requirements, undertake design tasks, apply configuration management to designs and associated documentation, whilst continuously monitoring the effectiveness of those activities for a given material state' (Ministry of Defence, *MOD Ship Safety Management: Issue 3, Part 1: Policy*, Ship Safety Management Office, July 2004b).

tained a core staff that held a variety of technical expertise to maintain design authority and to perform the top-level design work. The MOD hired a subcontractor, Yarrow–Admiralty Research Department (Y-ARD), which provided significant design support. It also hired a shipbuilder, Vickers Shipbuilding and Engineering Limited (VSEL), to perform the detailed design work and to build the submarine. MOD staff stationed at Barrow were heavily involved in production oversight, test, and acceptance. Various subcontractors designed systems and subsystems on the submarine. For example, Rolls-Royce and Associates designed the nuclear steam-raising plant (NSRP).

The manner in which the MOD managed its nuclear submarine acquisition programmes was revolutionised by a major cultural change—beginning in the 1980s and continuing through the 1990s—regarding the issue of the appropriate role of the government. Two successive Conservative administrations took a different philosophy to governing than had prevailed to date. That philosophy included a call for an overall decrease in the scale and scope of government. Industry was viewed as the rightful owner of many of the tasks that government had been performing. For example, the administrative infrastructure used to manage weapon system acquisition was often criticised as being large and costly—and inappropriate. All of this put pressure on the MOD to reduce staffing levels, which led to the elimination of whole organisations that formerly played a major role in submarine acquisition.

These changes resulted in the MOD following a dramatically different model for the Astute-class acquisition than for any previous class of nuclear submarine. Rather than being heavily involved in every stage of acquisition, the new approach involved, as much as possible, the outsourcing of work to a prime contractor. For the Astute contract, this included giving design authority to the prime contractor, BAE Systems. Because of a loss of talent and strict adherence to an arm's-length relationship policy at the outset of the programme, the MOD did not provide full support to the shipbuilder, which had been given greater authority than any previous shipbuilder over the highly complex task of producing a nuclear-powered submarine.

Although not the only cause, the new management model for Astute contributed to major cost and schedule problems with the programme. Currently, the programme is more than three years behind the schedule set during its initial stages and is approximately £1 billion over the initially approved cost.

Research Objectives and Approach

The loss of internal MOD talent, coupled with a decreased level of nuclear submarine procurement, raises fundamental issues with respect to the roles and capabilities of the MOD as it seeks to effectively oversee the design and production of nuclear submarines. The prior model of a talent-laden MOD presiding over an industry high on the nuclear submarine production learning curve served well in its time but was costly and given to micromanagement by the MOD. In the initial approach to Astute-class acquisition, however, the MOD's management capabilities had been reduced to a point at which its ability to manage its risks and perform its responsibilities effectively was significantly affected. There has since been some addition of capabilities that have made the MOD more effective at addressing problems in the programme. However, the MOD needs more action to arrive at a capability model that is best for the low production-rate environment.

Given the various advantages and disadvantages of the vastly different management models used by the MOD in managing nuclear submarine design and construction, the Attack Submarines Integrated Project Team (ASM-IPT) asked the RAND Corporation to address the following questions:

- What roles and functions should the MOD perform if it is to be a smart and engaged buyer of nuclear submarines?
- What management structure and level of resources are needed to perform those roles and functions?

- How could the MOD transition its workforce to the desired end state?

To address these questions, RAND researchers examined the historical record on UK submarine procurement to learn lessons from experience with the two distinct approaches, the model used through the Vanguard class and the initial model used for the Astute class. They conducted a wide range of interviews with various people who are, or were, in different MOD organisations involved with submarine design and acquisition as well as with private organisations such as BAE Systems' Submarines Division at Barrow-in-Furness. The researchers also gathered information and insights from similar organisations in the United States. These included the US Navy's Program Executive Office for Submarines, and Electric Boat, a division of General Dynamics that designs and builds US submarines.

We drew on insights from these interviews and from the economics, economic sociology, and management literatures to develop an approach to procurement that should help resolve some of the past problems experienced in UK submarine acquisition. This alternative approach represents a middle ground between the hierarchical management approach that characterised procurement through the Vanguard class and the contractual arm's-length alternative that was used in the early years of the Astute-class programme. The Astute programme has in fact been moving in the recommended direction. The proposed 'network', or hybrid, approach occupying this middle ground is broadly supported by management 'best practices' literature on the benefits of intermediate structures, including partnerships, joint ventures, and the like. It also represents a step in the direction of the successful, less contractor-centric approach taken in US Navy submarine acquisition, although we do not adopt the latter as a model here. We offer a detailed description of what such a partnership model means, including the details of enhanced MOD engagement through all the submarine acquisition phases, and make suggestions on how current capabilities could be improved to support this new model.

Organisation of This Report

In Chapter Two, we describe how submarines were traditionally acquired in the United Kingdom and how that approach changed to its current organisation and how its interactions with the prime contractor have changed. We do this in the context of the different phases of a submarine acquisition programme, from requirements determination through construction and support of the in-service submarines. In Chapter Three, we describe how the ASM-IPT has evolved from its original formation to its current organisation and in terms of its interactions with the prime contractor. Chapter Four offers a model that should enable MOD to fulfil its responsibilities. It describes the specific MOD and industry roles and capabilities for the new approach, including any increases required in resource levels. Chapter Five discusses other issues in submarine acquisition as well as steps the MOD should take to develop and manage the workforce necessary to be an informed and engaged partner in nuclear submarine design and construction.

CHAPTER TWO

Two Different Approaches to Submarine Acquisition

In this chapter, we describe the evolution of the management models used by the MOD in its nuclear submarine programmes. We start with the very-involved role up through the Vanguard class. We then summarise the changes that occurred in the 1990s and the impact they had on weapon system acquisition in the United Kingdom. These changes led to a more hands-off role for the Astute-class acquisition programme.[1] We present the various models in the context of the major stages in a submarine's design, construction, and ownership life cycle.

Vanguard Acquisition Model[2]

In a major defence system's life cycle, the typical approach is to start with desired capabilities or needs and gradually work towards a realisable, producible alternative. This is a process involving the development of greater insight into what is wanted, how much it will cost, appropriate trade-offs between cost and capabilities, and finally the

---

[1] The Astute programme was caught between the two very different management models. It had its beginnings in the model used through the Vanguard class but was forced to evolve to fit the hands-off role adopted by the MOD by the time the initial contract was signed.

[2] Peter Duppa-Miller, Head of the Submarine Library, greatly facilitated our understanding of the various organisations and their roles during the acquisition of the Vanguard class of submarines.

specific design solution. The design solution is then converted into a detailed design, manufactured, accepted into service, and operated by the Royal Navy, hopefully providing the capabilities that were originally desired.

In submarine acquisition up through the Vanguard class, the MOD played a large role in all stages of the acquisition life cycle. The Naval Staff at Whitehall set the requirement for a new class of submarines. The Director General Submarines (DGSM) organisation formed a project team that generated the layout of the new submarine and defined the major equipments and systems to be included in the new class. This team took responsibility for the performance of the system and set the standards for design and construction. It relied heavily on naval engineering standards generated by technical organisations within the MOD. Groups on-site at the shipbuilder provided oversight of the construction process and integrated closely with other MOD organisations that were responsible for submarine acceptance. Overall, the MOD acted as design authority and prime contractor. This close involvement in the design and build of submarines resulted in significant technical resources residing within the MOD. The following subsections describe the organisational roles played by the MOD in the various phases of a submarine's life cycle through the Vanguard class.

Top-Level Requirements

The first step in any submarine programme is[3] to generate top-level requirements, which are derived from an assessment of the prospective mission needs and the concepts of operation of the new submarine. There may be a new threat, perhaps a new class of technically advanced enemy submarines, or some other reason for a new operational mission that existing classes of submarines cannot adequately perform. This leads to the generation of a new top-level requirement.

The combination of the top-level requirements, the prospective submarine missions, and the submarine concepts of operation

---

[3] Note that we often use present tense in this chapter, despite describing the Vanguard approach. We do so for practices that are still followed in a general sense.

together establish the design objectives. Generically, design objectives should include, at a minimum, three parts:

1. The submarine should perform its intended mission set.
2. The submarine design should be suitable for construction (within the resources available).
3. The submarine should be capable of employment as specified by its concept of operations, including safety of operations.[4]

The submarine capabilities desired by the Royal Navy have been driven by the role of the UK military in the geopolitical arena. UK submarine development and design are influenced by the country's role as a leader in NATO and the potential use of its submarines in associated conflict planning. For example, the Dreadnought class (laid down in 1959, launched in 1963) was developed in the context of UK participation in the defence of sea-lanes of communications. These were required to be kept open against a far greater number of Soviet submarines threatening Europe.[5]

Later classes of submarines, including Valiant, Swiftsure, and Trafalgar, were also driven by requirements responding to the Soviet threat during the Cold War. These requirements included the need to optimise quieting (to avoid the submarine's detection), improve sensing (to detect the submarine's opponent as early as possible), increase speed, and enhance combat system effectiveness.[6] These four capabilities were developed simultaneously to maximise the overall potential for antisubmarine warfare. As time progressed, additional capabilities, such as those to deliver mines and cruise missiles, were added to the boats.

---

[4] F. W. Lacroix, Robert W. Button, John R. Wise, and Stuart E. Johnson, *A Concept of Operations for a New Deep-Diving Submarine*, Santa Monica, Calif., USA: RAND Corporation, MR-1395-NAVY, 2001, offers an example of mission and concept of operation basis for submarine concept design.

[5] Norman Friedman, *Fifty-Year War: Conflict and Strategy in the Cold War*, Annapolis, Md., USA: Naval Institute Press, 1999, Chapter 18.

[6] Combat systems include sonar, fire control, and weapons employment complex (e.g., torpedo battery).

The distillation of desired capabilities into a contract for the construction of a new class of submarine was a many-staged process, led by the Procurement Executive organisation (which in 1999 became the Defence Procurement Agency [DPA]) in London.[7] The first step was drafting an outline staff target, describing the initial stages of a need as it begins to emerge. This then led to the development of a document called a staff target. The staff target identified the defence need in more detail and the likely capabilities that would be necessary to meet that need.

At the time, the Directorate of Operational Requirements (Sea) (DOR[Sea]) was responsible for drafting the staff target. DOR(Sea) was a predecessor organisation to the current maritime Directors of Equipment Capability (DECs) but was responsible only for operational requirements.[8] Within DOR(Sea), the Underwater Warfare Directorate comprised approximately 12 uniformed staff officers[9] who defined submarine operational requirements. These officers were assisted in the drafting of the staff target by other submarine experts in sister directorates of DOR(Sea) (e.g. operational combat systems, combat systems, and operational warfare,). DOR(Sea) also went outside the organisation to help write the content of the staff target, which was largely composed from a standard template. In many ways, the staff target was a predecessor document to today's User Requirements Document (URD).

Once written, the staff target was given to the DGSM in the Procurement Executive for further study and development. In 1997, the DGSM worked for the Deputy Chief Executive of the Procurement Executive (South of England) and was responsible for all submarine-related procurement activities. The DGSM included a

---

[7] The submarine part of the Procurement Executive was based in Bath.

[8] This differs from the DECs in the Equipment Capability Customer who are responsible for not only defining the operational requirement but also funding it.

[9] As of May 1997.

number of smaller directorates, each responsible for a specific area of underwater warfare. A selection of these included the following:[10]

- Director Nuclear Propulsion
- Director Submarines
- Director Combat Systems/Submarines
- Director Sonar
- Director Torpedoes and Weapons
- Director Naval Architecture
- Director Future Projects (Naval).

Up through the Vanguard class, the approved staff target triggered the funding for a variety of industry studies that examined the various new capabilities that were desired. These were comparative operations analyses and cost-benefit trade-off studies in which new potential submarine attributes were prioritised. The study outputs were drawn together to form the staff requirement, also known as the Navy staff requirement. The defence staff in London generated the staff target and staff requirement, with significant input from the operating flotilla and the intelligence community. The Procurement Executive would work with the Naval staff to ensure that the staff requirement was achievable and then develop more detailed specifications to enable a contract to start the actual work.

Initial Studies

After the initial top-level performance and mission requirements are defined, the designers begin a series of studies that lead to the development of the submarine specifications. Initially, *concept studies* aim to provide broad technical alternative solutions to operational needs. Gross submarine sizes, propulsion and other major system configurations, and mission equipment suites and layouts are proposed and adjusted to gain insight into alternative approaches to an overall submarine design that will meet the requirements. Concept studies

---

[10] This list is not inclusive but contains the primary directorates in the DGSM's area of responsibility.

can reflect alternative arrangements of the major ship mission systems, including torpedo tubes, control surfaces, masts and antennae, sonar systems, and the different arrangements of ballast tanks.

The desired outcome from these studies is a variety of submarine concepts that will all (in theory) satisfy the high-level requirements for the submarine. In doing so, the studies highlight questions and issues, such as materials selection or subsystem performance, which need to be studied in greater detail. Questions raised in the concept studies can then be addressed in more detailed feasibility studies.[11]

The next step is to undertake *feasibility studies* on the more promising concepts. These studies include engineering and cost analyses performed in adequate depth for the customer to understand the implications of proposed new technology for ship systems. Feasibility studies allow the general submarine concept to progress into a better-understood and finely detailed ship proposal. For example, the feasibility analyses initially estimate the important 'budgets' such as those for weight and power. These are important because they set bounds on the design if the decision is to proceed to that phase. Feasibility studies also provide the opportunity to fully characterise the operational and cost impacts of new technology under consideration.

Once the decision is made to proceed with a particular design, the challenge then is to translate the selected submarine concept into language suitable to support the contract for the ship detailed design and build process. It is important that the *specifications* be detailed enough that both the customer (MOD) and the contractor have an identical understanding of the submarine and its systems before entering into a construction contract. Effort expended at this stage can prevent misunderstandings and problems later on during construction and acceptance.

While they derive in part from the new concept and feasibility studies, the full ship's specifications generally include information from previous classes of the submarine. For example, as a successor

---

[11] Roy Burcher, Louis J. Rydill, I. Dyer, R. Eatock Taylor, J. N. Newman, and W. G. Price, *Concepts in Submarine Design*, Cambridge Ocean Technology Series, Cambridge, England, 1995.

class, the Trafalgar's specifications followed the Swiftsure model, avoiding unnecessary modification. The ship's specification also included items addressing 'sore thumbs' or feedback from operational deficiencies from the prior classes. Also, military capability deficiencies were addressed.

The time from initial requirements generation to submarine delivery was quite lengthy, perhaps ten years. It may have taken two years just to move from an outline staff target to the staff target, an additional two years to develop the staff requirement, and another year to write the main contract. By the time of the first in class's completion, the staff requirement might have been out of date. The threat that the staff requirement was designed to meet might have changed, technology improved, or the costs of ship maintenance increased.

To deal with these potential changes, a Class Policy Document was created, which would articulate the requirement as seen by the defence staff in London to improve capability and better manage maintenance costs. This document was started at the same time as the initial staff requirement and typically spawned a number of additional staff requirements to improve individual aspects of the submarine. The first ship in the class was retrofitted with the changes, which were built into subsequent units. For example, the Swiftsure and Trafalgar ended up with about 700 change notices each.

The shipbuilder at Barrow, the lead nuclear submarine–building shipyard, played very little role in the studies process. Hundreds of engineers and designers in the Procurement Executive developed the initial system and arrangement drawings that formed the basis of the contract to build the submarine. However, the MOD did not have a monopoly on design expertise. The shipbuilder also employed many designers for the next stage of the process.

Detailed Design

After the ship contract specification is completed and the contract signed, the builder can start detailed design, translating the specification, systems diagrams, and arrangement drawings into design products (detailed assembly drawings, computer-aided machine instructions, etc.) that can be used on the shop floor and on the waterfront to build the ship. This task is remarkably complex, requiring the integration of more than 200 ship systems. As a process, detailed design demands aggressive management to ensure that drawing output is both sequenced properly and paced to support the planned production programme.

During detailed design, the designer completes the necessary modelling work to ensure that the ship's systems, such as those related to steering and diving, are properly designed and arranged. Flotilla input and experience is particularly valuable when designing the control station arrangement and control system operating features. Flotilla input is also valuable when considering how to best insert and apply new technology.

The contract is a 'living' document—that is, it changes throughout the lengthy submarine build. Some specifications are fixed, such as the dimensions of the pressure hull, but systems as significant as the main sonar can be updated. The second ship in the class would typically include perhaps 100 of the 700 change notices for the first of class. The additional 600 would be retrofitted, and perhaps another new 100 staff requirements would be added. The process meant that, while all members of a single submarine class might look identical from outside, the details could be very different, and later ships would represent improvements over earlier ones.

The Procurement Executive acted as a prime contractor and provided a range of systems, including the NSRP, to the shipbuilder as government-furnished equipment. The Procurement Executive also served as the integrator of the various systems into the submarine. To support its role in the acquisition process, there was significant technical expertise in the Procurement Executive, with a great depth and breadth of experience. Expertise was located at a number of sites in the United Kingdom and included members of the Royal Corps of Naval Constructors and engineering officers of the Royal Navy. The platform requirements organisation was located at Bath. Combat system responsibilities were split between Portland and Portsdown, so that they could be collocated with the relevant bits of the research and development organisation. The Procurement Executive held the design authority and performed top-level design in Bath with the help of support contractors, including Y-ARD.

Construction

The overall objective of the construction process is to achieve the contracted construction quality on schedule and at minimum cost. As the detailed design is being completed, material requirements are identified and sourced for those items that will require a long 'lead time' for the suppliers to construct and deliver. Typical amongst these are components of the NSRP and the sonar and weapons fire control systems, as well as hull steel or other special application metal alloys.

The shipbuilder performed detailed design and construction, but had a great deal of oversight from the MOD. The DGSM project team had significant design and integration expertise. This team, as well as the shipbuilder, drew on various equipment related teams (e.g., hydraulic pumps) and specialist sections within the MOD. These specialists came from the Chief Naval Architect, the Director General Marine Equipment, the Director General Underwater Weapons, and the Director General Surface Weapons. The specialist areas encompassed significant knowledge and experience, but the integration of their inputs (at times, requirements) could be difficult.

The Procurement Executive had a representative at the shipyard in Barrow called the Principal Naval Overseer (PNO), whose staff numbered about 50 people. Throughout the 1960s, 1970s, and 1980s, the MOD gained confidence in the acceptability of the submarines from the insight of the PNO. With the PNO present throughout production, quality assurance, test, and validation, the MOD had in place a mechanism to independently ensure that its nuclear submarines were built to contract specification, built as designed, and ready for final acceptance. The PNO could approve small changes, such as those related to material and layout. The PNO also provided deep insight into the schedule and monitored costs very closely, looking at every element on a daily or weekly basis.

Acceptance

Acceptance is a phased process designed to progress from the beginning of production through the delivery of the ship; it ends with the customer accepting the submarine. Acceptance is contingent on the customer's confidence that the submarine meets construction specifications and is therefore both safe to operate and appropriate for the Royal Navy's operational needs.

To achieve this confidence, the acceptance process incorporates an in-depth examination of all the systems on the ship. Starting at its lowest level, the process includes receipt and testing of the smallest system component upon its initial arrival in the shipyard. For instance, the builder is required by contract to maintain the pedigree of specified submarine components (e.g., controlled materials such as strength-specified corrosion resistant fasteners installed in sea water systems; pumps and valves that are tested, verified, and certified for proper material content as well as operational capability). Further system tests and inspections are conducted when the builder completes systems on board the submarine. The system tests could include, for example, previously certified pumps and valves, along with newly certified intermediate piping that completes the flow paths between major components previously untested.

Appropriate processes are followed for component and system test, verification, certification, and acceptance. These processes cover all the systems on board the ship, including those that span the submarine, and are developed during the submarine design process, often in conjunction with the subsystem vendors.

Contemporaneous with completion of the ship's construction, all ship's system, component, and subsystem testing is completed. This testing was performed by the shipyard's Dockside Test Organisation, which incorporated individual joint test groups consisting of members from the shipyard, ship force, and the MOD. These groups prepared the test agenda and acceptance criteria, executed the tests, and documented the results. Because of the nature of a submarine, system testing and verification are designed to be thorough and exhaustive. Overall testing on the ship includes, for example, nondestructive inspection of all pressure hull penetrations and closure welds, nondestructive testing of system boundaries to sea pressure, and other pressure hull integrity and sea-worthiness checks. The ultimate level of system testing is the ship's sea trials, during which all systems operate as an integrated whole as the submarine is tested at sea against its top-level requirements, such as speed and turning radius.

As the submarine neared completion and drew closer to delivery, the Captain Submarine Acceptance (CSMA) staff would carry out intermediate inspections and, eventually, the final inspection. The staff looked at every part of the submarine, examined all the test documentation to make sure that all elements had been signed off, and generally took on the responsibility to ensure that overall quality was sufficient to enable the ship to go to sea.

Reportedly, half a dozen or so people from the CSMA's staff took three days to perform a static inspection of the parts of the ship. At that point, the submarine was ready to go to sea for trials. After the trials had achieved results satisfactory to the many Procurement Executive and CSMA experts, the CSMA authorised acceptance. The entire acceptance process was characterized by visibility to all interested parties, whether within the government or the shipbuilder.

Support

After the submarine is accepted and enters the fleet, it needs to be supported through its lifespan. This support includes a variety of tasks with widely differing levels of complexity, from the ongoing supply of spare parts or consumable supplies to regular servicing and maintenance, submarine refits, and refuelling operations.

A submarine refit involves the removal of a submarine from active-duty service for several purposes. Its pressure hull is examined for integrity. Other mechanical and electrical services are checked and repaired or replaced if required. Updated technology can be incorporated onto the vessel to improve operational performance, enhance overall submarine safety, or manage obsolescence. The final step is a series of tests to ensure that each system and the submarine as a whole are operating properly. Refuelling is a much more complex operation that involves replacing the ship's nuclear core. Current UK classes of submarines were designed to have one refuelling operation during the life of the ship, but the Astute-class reactor core is designed to last through the life of the submarine.

During the support phase, every effort is made to ensure that lessons learned from any one class are transferred through the entire submarine fleet for effectiveness and that fleetwide management practices are used for efficiency.

The Naval Support Command had overall responsibility for the management of Royal Navy assets, including bases, ships, and submarines. Within the Naval Support Command, the Ships Support Agency had direct management responsibility for maintenance and modernisation actions for naval vessels.[12] The Director of In-Service Submarines provided those capabilities for the submarine fleet. The actual work was carried out in the Royal Dockyards.

Evolving to a New Approach

We next describe the transition to a new market model for the acquisition of submarines, of which the Astute class was the first case.

Factors That Led to the Change

The model that the UK government used up through the Vanguard class to buy major weapon systems had both benefits and costs. The government did have a lot of control over the process, as well as much insight into what happened at the shipyard. The internal structure meant that the government had a wide range of skills, capabilities, and experience drawn from across the flotilla of submarines with which it could manage the process. The management model through Vanguard was effective but not necessarily efficient.

Although the Vanguard project was delivered within budget, one major criticism of the historical approach was its expense. The

---

[12] The Warship Support Agency was formed in April 2001 by merging the majority of the Naval Bases and Supply Agency and the Ships Support Agency, both of which were part of the Naval Support Command.

administrative infrastructure and in-house technical staff used to manage weapon acquisition were large. Concerns arose as to whether the bureaucratic transaction costs had grown to exceed the value added by the infrastructure. In concert with the philosophy that the MOD should bear the risks of submarine acquisition, shipbuilding contract prices were based on cost and provided little incentive for the shipbuilder to invest in efficiency improvements over the long run. Costs also increased with requirements growth over the long acquisition period necessitated by the cumbersome sequence of stages and milestone products.

Another criticism reportedly was that the government, notwithstanding the number of engineers on staff, was relatively conservative in accepting production innovations. The government did invest in research and development of many aspects of submarine construction and directly invested in infrastructure, such as the Devonshire Dock Hall and shiplift at Barrow. However, it was believed that by giving the shipbuilder the freedom and incentive to innovate, production improvements would be more rapid.

But these specific criticisms did not directly drive the change to a new model. The criticisms of how weapon systems were purchased were grounded in a broader cultural change in the 1980s that led to a shift in government philosophy. Prime Minister Margaret Thatcher was elected on a platform that promised to decrease the size of government and the number of civil servants. This eventually put pressure on the MOD to cut its staffing.

Entire organisations were eliminated, including the PNO and the CSMA, whose functions have since proven to be necessary for oversight of a programme as complex as the design and build of a nuclear submarine. Recruiting for future needs was also curtailed. For example, the Royal Corps of Naval Constructors, one of the key organisations in the design support and build oversight of naval defence systems, stopped recruiting new members. Today, it is a much smaller organisation.

Closely linked with the idea that government was too large was the idea that it was engaged in activities that were more appropriately situated in industry. The free-market philosophy espoused by the

Conservative government held that competitive pressures would force industry to be more efficient and effective.[13] Since government did not have these pressures, it would not improve in the same way. Industry also had more skills and more flexibility. Nuclear submarines up through the Vanguard class were thought to be expensive because the shipbuilder did not have incentive to improve in the cost-based contracting environment. Thus, competition was seen as helping resolve these problems. Note that any one concern might have been solved differently by itself. For instance, cost-based contracting might have been improved through the addition of incentives. However, in the broad view, a new approach was seen as the solution to problems identified in acquisition programmes, including those for submarines.

Beginnings of the Astute Programme

Submarine classes through Trafalgar and Vanguard came at intervals that sustained a steady stream of production and product improvement. Preparation for the Astute class originally followed the same timescale, with the original studies being conducted in the 1980s, when Trafalgar and Vanguard were in production. The SSNOZ (the original name for the class) was to represent a major change in capability from the Trafalgar class, which by contrast had been an evolutionary upgrade from the Swiftsure class. The SSNOZ became SSN20, whose features were to include an improved nuclear propulsion plant, an integrated sonar suite, a large increase in firepower, new combat systems, a larger pressure hull with new steel, increased stealth characteristics, and control surfaces modified for enhanced agility. The staff target was developed, and within an additional four years the staff requirement was developed and agreed upon within the MOD.

This work coincided with a change in the geopolitical strategic climate. The fall of the Berlin Wall and the breakup of the Soviet

---

[13] This philosophy also affected the in-service support of submarines. The Royal Dockyards were privatised, and the new organisations were encouraged to compete for submarine maintenance and refit work.

Union meant that the threat confronting the United Kingdom had dramatically altered. This called into question the plan for the SSN20, which began to be perceived as too expensive and unnecessary. Submarine work essentially stopped. A compromise resulted in the MOD's spending some remaining concept development funds on industry studies to generate concepts for a less ambitious 'Batch 2 Trafalgar', with a Vanguard-class NSRP.

In line with the principles of the Thatcher government, the idea of industry being the source of 'prime contractor' management expertise began to take hold, replacing the idea of industry simply building the ships. It was also viewed that increased competition in the defence industry would help solve the problem of cost growth while encouraging more innovative solutions to meet strategic requirements.

The studies phase for the Batch 2 Trafalgar began around 1992. Four potential contractors took part in these studies, including VSEL, GEC Marconi, Rolls-Royce and Associates, and BAE Systems. VSEL had the most experience with submarines, with all but three of the United Kingdom's nuclear boats having been built at its Barrow yard. Rolls-Royce had traditionally been responsible for the NSRP on all classes of nuclear ships. The other two contractors had less experience. However, the companies all indicated that they could produce a submarine to meet a reference design that fulfilled the Batch 2 Trafalgar staff requirement. A draft invitation to tender was distributed to all of the companies in October 1993, with the final version sent in July 1994. At this time, the Secretary of State announced to Parliament that there would be enough money to buy at least three of the next class of submarine.

Despite the trend towards less direct MOD involvement, the design was very closely specified. The invitation to tender contained some 2,000 commercial requirements and about 12,000 system technical requirements. Two completed tenders were received by MOD in June 1995, since by this time GEC Marconi was working with Rolls-Royce, and VSEL with BAE Systems. GEC Marconi was chosen as the preferred contractor in December 1995.

GEC Marconi was selected because its price was substantially less than that of VSEL. Its production plan was also perceived as innovative, relying on a modular approach in which large assemblies were put together off the boat and then slid in. The plan involved constructing the submarine in sections at shipyards in the northeast of England, which had lost other work and would work for reduced cost to provide jobs. The boat was to be assembled and fuelled at one of the refitting yards. However, the GEC tender was characterised by one interviewee as not reflecting an understanding of detailed design and ship construction.

By contrast, the VSEL tender was reportedly more conservative with regard to cost risks. This tender recognised the fact that the supplier base had not designed and manufactured parts for submarines for many years and that costs would therefore be higher. The plan was to do the work, as usual, at the Barrow shipyard. The VSEL approach was characterised in one interview conducted during the course of this research as 'an expensive and dull design'. But the design met the requirement, and subsequent events have shown that this tender had more cost realism than that of its competitor.

The prime contract was signed in March 1997. By this time, GEC had bought VSEL and decided to build the submarine at Barrow. GEC sold its shipbuilding business to BAE in 1999. Hidden within this industry churn is the fact that at the same time, the MOD, having decided to reduce its role in designing and building submarines, began losing its design talent and design and build oversight capability.

The UK government worked with GEC on the acquisition approach before the contract was signed. The government followed the No Acceptable Price, No Contract approach. This policy required that the government agree to a price within affordability constraints prior to placing a contract. In doing this, the government and the company overestimated the cost reductions expected from modern design tools and manufacturing processes.

The chosen contract approach was also billed as transferring the management of risks to the prime contractor, which helped justify less MOD oversight. The MOD believed that the prime contractor could better manage the shipbuilder and that in the future it might be possible for the prime contractor to compete work amongst various shipbuilders. To that end, the contract specified separation between the prime contractor, then GEC Marconi (now BAE Systems), and the shipbuilder at Barrow.

The contract was structured on a fixed-price basis, with significant penalties if the contractor failed to meet its promises. In the spirit of transferring the management of all risks to industry, the contractor received more authority, including formal design authority, over the Astute-class programme, with less government intervention and interference. Also, assuming the contractor could reduce in-service costs through design trade-offs, the contract contained eight boat-years of contractor logistics support,[14] a radical change from previous submarine contracts. The contractor would have the ability to make design trade-offs as long as the result met the specifications in the contract. To further control costs, the contractor agreed to provide three boats with the same configuration rather than ones that individually incorporated successive changes, as for previous classes.

However, the MOD could not pass full responsibility to the prime contractor. As part of making the prime contractor responsible for the totality of the project, Rolls-Royce, the NSRP design authority, was specified by the Astute prime contract to be a subcontractor to the prime contractor. Because the prime contractor had no prior involvement in the design or procurement of components for the Pressurised Water Reactor 2 (PWR2) NSRP used in Astute, Clause 72N of the prime contract recognises the MOD as the authority ultimately responsible for nuclear safety. Moreover, the MOD, per Clause 71, also indemnifies the prime contractor with respect to the fissile material used as the energy source for the submarine.

With regard to the submarine itself, the MOD is responsible for the submarine not only after acceptance by the ASM IPT from the contractor (except for warranty or contracted support requirements) but also during construction, as Clause 16 of the prime contract makes the MOD responsible for loss, damage, and liability as if the MOD were the insurer under a marine insurance policy for ship con-

---

[14] The contractor was to provide logistics support for the first several years of Astute 1's service life and a smaller number of years for Astute 2 and 3, amounting to eight years in sum.

struction. Clause 61 of the Astute prime contract makes the MOD responsible as well for the cost of safety upgrades mandated by regulatory authorities. These prime contract terms mean that MOD involvement in the technical decisionmaking process for the Astute's design and build is necessary for the MOD to manage effectively the risks it has accepted for the Crown.

In hindsight, reducing contract oversight capability while retaining significant cost risk and ultimate responsibility for the most complex defence system built in the United Kingdom was detrimental to the government's interests. Those we interviewed suggested that more time, perhaps as much as six months, would have been useful to hammer out all the issues and to develop a better understanding of the requirements and the eventual price of the contract. However, it was clear that a change in government from the Conservative party to the Labour party would occur in the next election. The contract had to be signed before the general election process started, so the sitting government could not be accused of buying votes with the promise of defence industry jobs. The Labour government had signalled that if elected, it would conduct a lengthy strategic defence review, and there was concern that the resulting delay would negatively affect the shipbuilding industry, its cost structure, and other factors. The prime contractor and the government thus did not spend the time required to develop a mutual understanding of the contract requirements.

Moreover, much of the work that the government had done before had never been done by industry, and industry did not have the skills needed when the new contracting scheme was put into place. Furthermore, industry did not appear to understand the types and depths of skills that would be required.

As the submarine-related organisations in the Procurement Executive adapted to the radically different management approach, a second change in the acquisition of weapon systems was begun.

Smart Acquisition

In 1997, a Labour government was elected, replacing the Conservatives that had held power for almost two decades, spanning the leadership of Margaret Thatcher and John Major. The new government of Tony Blair had an agenda for change and modernising government, which included the operations of the MOD. A revised system for weapon systems acquisition called Smart Procurement (later changed to Smart Acquisition) was developed with the assistance of consultants McKinsey & Co.[15] A number of problems[16] were identified, including ineffective and underfunded early stages of the procurement process, too much reliance on technical specifications and competitive tendering, weak project management, too little delegation, key people spending too little time in post, low accountability, poor scrutiny, and insufficient incentives and penalties. The consultants recommended many changes, including a new acquisition cycle called the CADMID[17] approach, which replaced the Downey cycle. Other changes included clarification of the internal MOD customer-supplier relationship, incremental acquisition, an approach that considered through-life costs and issues, and personal accountability.[18] Finally, McKinsey recommended 'partnering to involve industry more closely in the development of operational requirements and equipment designs' and 'Integrated Project Teams (IPTs), consisting of all stakeholders and scientific staff as well as industry when competition allows, with a clear leader having authority to make trade-off decisions and who would retain responsibility for equipment after it had entered service'.[19]

The Smart Acquisition approach has offered many reasonable changes and fits into the broader cultural change regarding the reduced role of government that had been growing for the previous decade or so. However, we have no evidence that the analysis that led

---

[15] Bill Kincaid, *Dinosaur in Permafrost*, Walton-on-Thames, Surrey, England: TheSAURAS Ltd., 2002.

[16] Kincaid (2002), p. 4.

[17] The phases are concept, assessment, demonstration, manufacture, in-service, and disposal. They contain all the acquisition phases described in the previous chapter. See UK Ministry of Defence, *The Smart Acquisition Handbook,* Edition 5, Director General Smart Acquisition Secretariat, January 2004a.

[18] Kincaid (2002), p. 5.

[19] Kincaid (2002), pp. 5–6.

to Smart Acquisition was based on an understanding of the specific government context in which the DPA operated. It is also unclear that the analysis took into account the limitations of industry, including whether it had the skills to take on the responsibilities that the government wanted to outsource. It has also been criticised for not going far enough. For example, much of the focus on keeping key people on post has referred to IPT leaders rather than personnel within the IPT who hold key technical skills (who often must rotate to different IPTs for career broadening, which may degrade technical expertise over time).

Impact on the Astute Programme

The original contract for the Astute class was signed in March 1997, before Smart Acquisition became the official policy of the MOD. However, as the new approach changed how acquisition occurred, this inevitably spilled over into the Astute-class programme. Changes influenced by the Thatcher government and Smart Acquisition had transformed the landscape of the MOD. Gone were the Director General organisations as well as the PNO and the CSMA. The specialist groups were either eliminated or reduced in size (to later morph into the Naval Authorities). The DECs were formed to determine requirements and to oversee funding issues. Small but integrated IPTs were established to manage defence system procurements.

In some ways, the Astute class was a test case, although the acquisition strategy was backfit to Smart Acquisition. By 1999, the Director Submarines had moved to Abbey Wood and had incorporated an element of the DOR(Sea) community based there.[20] In addition, the Batch 2 Trafalgar programme had officially become the Astute programme. Figure 2.1 shows the organisation chart of Director Submarines as of June 1999.

---

[20] This group mostly conducted finance and secretariat functions.

Figure 2.1
Director Submarines Organisation, June 1999

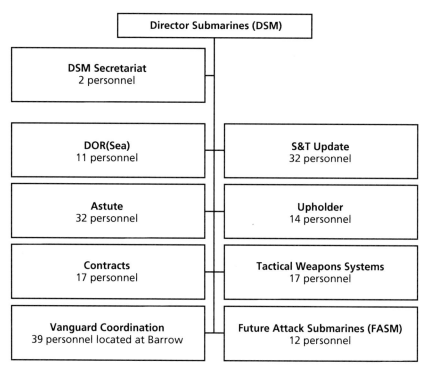

RAND *MG326/2-2.1*

Compared with two years earlier, the directorate had grown considerably to approximately 176 personnel, although the portions of the directorate that stayed (namely the contracts, science and technology [S&T] update, and Astute sections) remained approximately the same size. Additions to the directorate included Vanguard coordination responsibility,[21] the new Upholder-class programme, and the Future Attack Submarine (FASM) team, which moved to Direc-

---

[21] Vanguard maintenance functions moved to the Director In-Service Submarines in the Ship Support Agency.

tor Submarines from the old Director Nuclear Propulsion and Director Future Projects (Naval), where it was being developed in 1997.

This organisation was restructured under the Smart Acquisition philosophy to form the ASM-IPT. In July 2000, the ASM-IPT had 96 people, with a team leader, an assistant to the team leader, a requirements manager from the DEC, and four suborganisations— Finance and Business Development (8 people), Astute Class Procurement (43 people), S&T Class Upgrade (32 people), and Commercial (10 people). Note that during this early period, there were only two or three people from the IPT stationed at Barrow. The MOD's objective in the management of the contract was keep 'hands off' to allow the prime contractor the freedom to deliver, avoiding interference that would add or take back risk to the MOD.

A System Requirements Document (SRD) was developed based on existing information, including the system technical requirements.[22] The latter, however, were reduced in number by three-quarters as they were flowed to the SRD. Part of the reason for this was a push for performance-based contracting, but part was due to the recognition that a contract could not reasonably be written that contained all the necessary details.

Problems in the programme, including cost overruns and schedule delays, led to a review in 2002. The review concluded that both the MOD and industry had underestimated the difficulties inherent in transferring design authority to an inexperienced contractor. Also, the early benefits of computer-aided design were falling very short of expectations. In response, efforts were made to enhance MOD capabilities to become a more 'intelligent' customer. We discuss the current management approach and staffing of the ASM-IPT in the next chapter.

---

[22] The URD, a product of the DEC required under Smart Acquisition, is not yet complete.

CHAPTER THREE

Evolution of the Management Model for the Astute Programme

Towards the end of 2002, the Astute programme was more than three years late and several hundred million pounds over the maximum contract price. A contract modification was agreed upon that removed the cost ceiling from the prime contract and specified a maximum level of cost liability for the prime contractor. Incentives were rewritten to encourage the shipbuilder to reduce costs. Furthermore, the prime contractor was integrated with (instead of being separated from) the shipbuilder. Most importantly, the contract recognised that risk management needed to be shared.

To help with the design effort, the ASM-IPT interacted with the US Naval Sea Systems Command and secured the help of Electric Boat, which designs and builds US submarines. The IPT realised that a more hands-on approach was needed and strengthened its interactions with the shipbuilder at Barrow.[1] In line with the increased responsibility and risk management taken on by the IPT, its staffing level increased from 96 in 2000 to the present level of nearly 130. A large part of this growth was at Barrow, where the contingent grew from the original two to a planned complement of more than 30.

The previous chapter described how the requirements, initial studies, and detailed design phases of the programme were caught in

---

[1] Muir Macdonald, 'How Team Made an Astute Move', in UK Ministry of Defence, *Excellence in Defence Procurement: Equipping the Armed Forces*, Defence Procurement Agency, 2004, provides an informative overview of the changes that have occurred in the management of the Astute programme.

the transition between the very hands-on approach up through the Vanguard programme and the more arm's-length philosophy of the 1990s. In this chapter, we discuss how the management of the Astute programme has evolved by describing the MOD's current approach to design, construction, acceptance, and support phases of the programme. We start with the technical decisionmaking process, since it affects the design and construction phases.[2]

MOD Involvement in Design and Construction

There are about 2,500 contractual technical requirements for which design and/or construction acceptance criteria are being prepared. If, as it evolves, the design of the submarine requires a change to any of these requirements, a deviation requiring MOD approval is processed. This deviation process goes through the ASM-IPT, which can handle the matter itself or seek expert outside technical advice.

If any aspect of the submarine detailed design or as-built condition does not meet the design intent, a concession is requested so that the prime contractor design authority can determine whether the as-built or designed condition meets the requirements. If so, the design authority can approve the concession; if not, corrective action becomes necessary. The concession process is conducted within the prime contractor's organisation, though the MOD is kept apprised of the whole process. The MOD can challenge any part of the process if it considers higher-level contractor or MOD attention is required.

The ASM-IPT includes 'requirement owners' who have SRD fulfilment and compliance assigned to them. They work closely with the prime contractor, on-site ASM-IPT members, MOD technical experts, and others to ensure that the system requirements will be met in the detailed design and build. While these 'requirement owners' have access to the MOD's technical experts in the DPA's Sea Technology Group (STG), the Equipment IPTs, and, by contract if neces-

---

[2] Note that technical decisionmaking during design and construction also affects the acceptance process.

sary, with outside experts in organisations such as Qinetiq, there is no formal requirement to use them. As discussed below, however, there is an informal, yet compelling, reason for the IPT 'requirements owners' to seek advice and concurrence from the technical experts in the STG concerning safety issues early in the design and build process.

Responsibility for safety within the MOD rests with the Secretary of State for Defence. With regard to sea systems, this responsibility has been delegated to the Controller of the Navy as Chairman of the Ship Safety Board. The Ship Safety Board's policy for safety management is set out in Joint Service Publication (JSP) 430, *MOD Ship Safety Management*. JSP 430 requires that the duty holder (for Astute, the ASM-IPT leader) ensures relevant safety requirements are met and that safety risk is reduced to a level that is broadly acceptable, or tolerable and as low as reasonably practicable.[3]

The duty holder must demonstrate that he has met his responsibilities by preparing a safety case, which JSP 430 defines as 'A structured argument, supported by a body of evidence that provides a compelling, comprehensive and valid case that a ship or equipment is safe for a given application in a given operating environment'. The duty holder must also commission an independent safety auditor to confirm that the safety regime is in accordance with policy and that the safety case is comprehensive.

A major element of a ship safety case is the regulation of key hazard areas by the Naval Authorities, who are independent of the duty holder and are authorised by the Ship Safety Board to regulate seven specific hazard areas where there is significant danger to peoples' lives, loss or severe damage to the platform, or significant damage to the environment. For Astute, these specific hazard areas are submarine stability, submarine structural strength, atmosphere control, manoeuvring and control, watertight integrity, shipborne explosives, and fire safety. Five of the Naval Authorities reside organisationally within the STG, and the other two reside within the Defence Logistics Organisation's (DLO's) Warship Support Agency

---

[3] UK Ministry of Defence (2004b), p. 5-9.

(WSA). The Naval Authorities issue certificates of safety based on audits of submissions from the duty holder. These certificates of safety can be considered the building blocks of the whole ship safety case, confirming compliance with standards and criteria.

A similar process is followed by the Nuclear Propulsion Integrated Project Team (NP-IPT), which takes the lead in ensuring, auditing, and documenting the safety of the NSRP for which the NP-IPT leader is the duty holder for Astute submarines. JSP 518, *Regulation of the Naval Nuclear Propulsion Programme*, governs in this area.

With regard to areas outside the safety case and the purview of the Naval Authorities, the DLO's Maritime Commissioning Trials and Assessment (MCTA) organisation provides additional oversight of the product of ASM-IPT technical decisions. The MCTA acts as an independent auditor and helps provide assurance to the DEC's Capability Working Group (CWG) that the maritime defence system meets the requirements established by the DEC. Additionally, the ship's crew who stand by during the latter part of the build period and operate the ship and its systems during the trials period can report their concerns up their organisational chain (independent of the ASM-IPT and even the DPA) to the fleet (Customer 2), which also has a seat on the CWG.

The prime contractor is required to have a quality assurance system in accordance with ISO (International Organization for Standardization) 9000, which requires third-party independent accreditation. The DPA also has a Quality Assurance Field Force, independent of the ASM-IPT, which audits the prime contractor's quality assurance processes and, more specifically, the Astute first-level (submergence safety) programme.

Effectiveness and Sufficiency of MOD Involvement

In sum, the ASM-IPT is the technical decisionmaker for the MOD and continuously ensures that the design and build process meets contract requirements.

The ASM-IPT is not alone in the MOD's Astute submarine technical decisionmaking process, nor is it the sole MOD organisation with formal responsibilities in this area. The requirement for an independent safety auditor to endorse the safety case, where appropriate, and the certificates required from the Naval Authorities provide independent oversight of the product of ASM-IPT technical decisions in the very areas where MOD has retained significant risk (i.e., potential loss of life, loss or damage to the ship, or severe damage to the environment). The DEC CWG ensures that the IPT presents a submarine that meets all of its requirements. The fleet has a member on the CWG who ensures the crew operating the ship through the trials period has an input independent of the IPT. The CWG relies not only on the IPT leader but also on the Naval Authorities, the independent safety auditor, the MCTA, and the fleet.

At the outset of the Astute programme, only two ASM-IPT persons were on site at the building location in Barrow-in-Furness. Today, there are more than 20, with plans for adding more. The magnitude of the safety and other risks retained by the MOD, coupled with the volume of work remaining to test and document acceptance of about 2,500 requirements, will likely stress the existing ASM-IPT organisation (at both Abbey Wood and Barrow-in-Furness) over the next few years. Simply because the ASM-IPT is now keeping up with acceptance-related demands, does not mean that it is staffed to meet an ever-increasing workload.

The delegations of authority, along with the provision of independent oversight of prime contractor and ASM-IPT technical decisions by multiple organizations, appear to afford the MOD with appropriate tools to manage Crown interests on the Astute design and build programme. Active engagement and oversight of the prime contractor at a level at least as much as provided by ASM-IPT today appear necessary to ensure effective and sufficient management of the risks retained by the MOD.

Efficiency of MOD Involvement

The Naval Authorities currently exercise their functions at the end of the long design and build programme. Ship conditions that require alteration to satisfy regulatory requirements at this stage can be quite expensive to rectify. Early involvement of the Naval Authorities could obviate the need for expensive changes if the unsatisfactory condition is realised prior to fabrication. However, there is no requirement for consultation with the Naval Authorities in the early stages of the programme.

As noted earlier, there is an informal incentive that compels the prime contractor and the ASM-IPT to consult with the experts in the STG (and the DLO as appropriate) early to avoid late changes required by the STG when it later acts as the Naval Authority on the specific hazard area. For this informal process to work effectively, the prime contractor designers and ASM-IPT personnel must not only know what the concerns of the Naval Authorities are today but also what they are likely to be years later when the safety certificates must be executed to allow operations. A formal requirement for the Naval Authorities to be involved in the early stages of a submarine programme could alleviate the potential for late, expensive design and construction changes.

MOD Involvement in Acceptance[4]

Acceptance of the submarine is a multistage process conducted by the MOD that confirms whether the submarine meets both contract specifications and the user's military capability needs. As such, it is more appropriate to think of three distinct but equally important stages of acceptance:

- **Contract acceptance:** Acceptance of the submarine by the IPT from the contractor.
- **System acceptance:** Acceptance of the submarine's systems by the wider MOD from the IPT.
- **In-service date:** Acceptance of the submarine into service by the wider MOD.

---

[4] Much of the information contained in this section is taken from the official MOD Acquisition Management System guidance on acceptance (UK Ministry of Defence, *Acceptance and the Achievement of New Military Capability*, no date).

When the MOD accepts the submarine from the contractor, it is confirming that the contractor has fulfilled the conditions of the contract (currently represented by approximately 2,500 system requirements that are laid out in the SRD). This allows change of legal ownership of the submarine along with a contract specified payment to the contractor. The IPT leader is the acceptance authority for this event.

Likewise, when the MOD accepts the submarine (systems and ship) from the IPT into service, it is confirming that the submarine satisfies the system (system acceptance) and user (in-service acceptance) requirements laid down by the equipment capability customer (ECC), i.e., the DEC. In theory, the URD serves as the benchmark to ensure that the into-service requirements are met. The DEC is the acceptance authority for the MOD for this event.

It is in the IPT's interest to ensure that these acceptance events happen as close together as possible, or are tied together in some way, to avoid the difficult situation of having accepted a submarine from the prime contractor that does not meet the Royal Navy's needs. The IPT and DEC thus form a strategy that identifies how the acceptance process will demonstrate the satisfaction of the ECC's user requirements. This strategy then guides the IPT and the contractor in creating an integrated test, evaluation, and acceptance (ITEA) plan, which ensures that the contractor will be able to demonstrate that the various system requirements of the submarine are met and verifiable. According to the MOD's Acquisition Management System:

> The ITEA Plan details how the cumulative evidence required to demonstrate the achievement of all user and system requirements, and any other contractual commitments, is to be obtained, including, where necessary, the use of independent advice. It will also cover those T&E [test and evaluation] activities required to inform other processes such as the development of tactics and procedures as well as the definition of safe operating environments. The collection and collation of T&E data is a through-life process which must be managed effectively by the IPT.

The ITEA plan addresses the performance dimensions of the capability; it does not address cost or schedule adherence. It is the primary mechanism used by the MOD to verify and validate that the submarine (or other appropriate equipment) meets contractual and user requirements. The verification aspects of the ITEA plan ask, 'Did we build the system right?' These aspects confirm whether the system complies with the SRD and provide the evidence that allows the IPT leader and the DEC to conduct both the contract and system acceptance of the submarine.

The validation aspects of the ITEA plan ask, 'Did we build the right system?' These aspects provide the evidence that allows the DEC to confirm whether the submarine will provide the military capability required by the URD. Verification and validation together give the MOD the confidence that the equipment will perform to its required capability.

Because the design and production cycle is so lengthy for a nuclear submarine, it is likely that some type of progressive acceptance system will be used. By accepting portions of the contract as soon as the criteria are met, the IPT leader not only verifies the equipment meets its requirement but also mitigates risk: should the equipment prove substandard, it can then be corrected at a much lower time and cost penalty than would be the case if acceptance were postponed until the end of production. When the IPT leader is satisfied that all contractual requirements have been met, he then accepts the submarine "off contract" from the prime contractor. At this point, the DEC can formally initiate system acceptance of the equipment.

As the in-service acceptance authority, the DEC is responsible for ensuring that each key element of military capability meets the MOD's user requirements. To manage this, the DEC establishes and chairs the CWG(Acceptance), which defines the requirements of each of the key element contributors. This group comprises the major stakeholders of the Astute programme—for example, the ASM-IPT leader, the DEC, the fleet, and the MCTA are all represented on the CWG(Acceptance). To formalise the contributor requirements, the DEC establishes customer supplier agreements (CSAs) that specify the outputs required from each stakeholder involved in the acceptance process. The CWG(Acceptance) monitors these outputs and is able to determine when the overall military capability is met, and then the submarine is accepted into service. In the case of Astute, the ASM-IPT leader is charged by the DEC with collecting all of the evidence showing that military capability has been achieved (through the delivery of the CSAs) and presenting it to the CWG(Acceptance) for approval by the DEC.[5]

The DEC also has access to the MCTA group to assist with testing and verification issues. The MCTA was formed around the advent of Smart Acquisition from members of a number of now-defunct testing and acceptance groups, including CMSA. The MCTA currently has on the order of 150 people and provides independent acceptance advice on both new build ships and those coming out of refit. The MCTA has no role in the Astute prime contract and thus does not have the authority to conduct tests and trials, but it will witness them as an independent expert.[6] When witnessing trials, the MCTA follows BR4050, *Instructions for the Conduct of Naval Weapons Inspections and Trials*. While the MCTA is also a free source of technical, acceptance, and testing advice to all naval IPTs, there is no explicit requirement for these IPTs to use MCTA. In the case of Astute, the MCTA supplies inputs to the ASM-IPT and augments the capability and technical expertise of the IPT's acceptance personnel in Barrow. Captain MCTA also is a formal member of the DEC's CWG(Acceptance) and is available to provide advice where appropriate.

Strengths and Weaknesses of Current Acceptance System

The current acceptance process and plan offers a mix of strengths and potential weaknesses to the MOD, although we would note that the advantages outweigh the disadvantages. On the positive side, the

---

[5] As the acceptance authority.

[6] The MCTA observation team size will vary from one to five members, depending on the type of trial, with weapons testing involving the largest team.

acceptance process is very integrated amongst the contractor, ASM-IPT, and the DEC, and the IPT has placed sufficient resources at Barrow to both verify and validate Astute construction. According to our interviews, these personnel are technical in background—which is a necessity given their roles. Both the IPT and the Royal Navy are involved in collaborating with BAE Systems to write testing standards and are on hand to observe those tests. The MCTA will also witness future tests to provide additional assurance.

However, the current system also offers some potential weaknesses as well. In the current construct, the DEC is almost wholly reliant on the IPT to provide acceptance data and has little internal capability to assess the inputs from the IPT. This potential problem is exacerbated further by the past lack of oversight within the IPT of design and construction activities occurring when the IPT had few people physically resident in Barrow to provide quality verification. Finally, the IPT will be focused on accepting the ship as fulfilling the contract, whereas the DEC will be interested primarily in whether the ship meets the user requirements. We believe most of these potential weaknesses to be theoretical at present, as the ASM-IPT is now actively and objectively managing the acceptance process. Furthermore, the IPT and the DEC have a formal coordination process under way to ensure consistency between contract and system acceptance.

While ASM-IPT manning is sufficient for now, we do have concerns over the number of suitably qualified and experienced personnel to handle the large volume of work in the next few years. The design and build adequacy of more than 2,000 system technical requirements will have to be reviewed and agreed on before Astute can be accepted from the contractor. Additionally, the role of the MCTA is not well defined in the contract and acceptance process. Clarification and formalisation of its responsibilities would help strengthen the MCTA's independent role.

MOD Involvement in Support

Maintenance and modernisation support for all submarines currently in service is managed by the Submarine Support IPT (SUB-IPT) within the DLO's WSA. The SUB-IPT also holds design authority for the Swiftsure, Trafalgar, and Vanguard classes. The refit and repair work is accomplished by either Devonport Management Limited (DML) at Her Majesty's Naval Base Devonport in Plymouth or Babcock Engineering Services (BES) at Her Majesty's Naval Base Clyde in Faslane, Scotland. The Submarine Support Management Group, consisting of DML, BES, and Rolls-Royce with subcontracts to such other firms as British Maritime Technology, assists the SUB-IPT in its management responsibilities for in-service submarines.

The original contract for the Astute programme included eight boat-years of support provided by the prime contractor, which would retain design authority. The renegotiated contract did not change this option. However, it is still uncertain whether BAE Systems will provide a cost for the contractor logistics support that is acceptable to the MOD. Whether BAE Systems will maintain design authority and play a role in the support of the Astute-class submarines once they enter service is still undecided. We will discuss the support of in-service Astute-class submarines in the next two chapters.

Conclusion

The proactive changes made by the ASM-IPT to become more involved with the contractor in the design and build of the Astute-class submarines is a very important step in the right direction. To foster this increased interaction, the IPT has increased its staffing levels at Abbey Wood, and especially at Barrow. We heartily endorse these changes, although we believe that the MOD could usefully go further, following a theoretical model we lay out in the next chapter.

CHAPTER FOUR

An Alternative Model for Managing the Acquisition of Nuclear Submarines

In this chapter, we describe an overarching model for managing acquisition of nuclear submarines. This model is based on literature from economics, organisational theory, sociology, and best business practices and was supported by many UK interviews identifying capability gaps in the current system of submarine acquisition. This model supports an approach to procurement in which the MOD is a fuller partner to industry. The organisational model represents, for the most part, an evolutionary change that builds from the capabilities of the MOD's existing component organisations. However, we do propose some changes to the current practices of the Astute programme and an increase in the staffing levels of various organisations.

The MOD owns certain aspects of risk in submarine acquisition that cannot be outsourced. Fully outsourcing the management of these risks is not possible, because if the burden of risks becomes too heavy for the prime contractor to bear, it will go out of business, leaving the MOD responsible for failing to satisfy the nation's need for a submarine. The MOD can manage these risks by accepting responsibility for them and developing appropriate capabilities to manage this responsibility. The management imperatives contributing to those capabilities ought, in our judgement, to be the following:

- Increase partnering with industry in submarine design and development, while increasing integration amongst MOD components to better manage this relationship.

- Increase ASM-IPT engagement in design and construction at Barrow to improve the MOD's understanding and oversight of the product and to increase support to the builder, thus raising the likelihood of the contractor's delivering acceptable ships on schedule and at projected cost.
- Resolve critical safety and technical issues early in the programme by requiring early involvement of the Naval Authorities.
- Ensure through-life ship safety, maintenance and postdelivery control of design intent, and the propagation of lessons learned across submarine classes, by transitioning design authority back from the prime contractor when the submarine enters service.

We first offer our perception of the risks the MOD faces in any weapon system acquisition programme and the responsibilities it must take on to manage these risks. We then present the new model by describing MOD roles at each stage of the submarine life cycle. Finally, we explain the roles of each organisation involved in submarine acquisition and how, if at all, the staffing of those organisations should change.

Where possible and applicable, we try to be specific about that last point—that is, we estimate the number and skills of personnel required in the MOD nuclear submarine–related organisations. This is a difficult endeavour at best. There are no objective, quantitative relationships that predict the number of personnel required to perform various duties in different situations. Therefore, subjectivity plays a major role in our estimates. In many organisations, such as the SUB-IPT and the NP-IPT, the workload will remain fairly constant with slight increases as a new class of submarine is designed, built, and introduced into the force structure. Other organisations, such as the ASM-IPT, will experience a sharp increase in workload as a new class enters design and initial production, followed by several years of slightly increasing or uniform workload as the new submarines are built, and then a steep decline in workload as the production run of the new class comes to an end. Without a new class of submarines, an organisation such as the ASM-IPT may need very few, if any, personnel resources.

Workload in the IPT responsible for the design and production of a new class of nuclear submarines will depend on many factors in addition to the phase of the overall acquisition process. Revolutionary programmes that introduce several new technologies will require more resources than an evolutionary programme that adopts technologies from existing classes. A contractor skilled in designing and producing nuclear submarines can reduce the burden on the MOD. Such skill comes from continuity as well as experience. That is, a gap in nuclear submarine design and production will increase the management workload at both the contractor and the MOD. A similar argument relates to the skill and experience of MOD personnel. The greater the experience of its staff, the fewer personnel resources the MOD will require.

In some cases, we give fairly specific manpower estimates, based on an increased workload resulting from a greater level of interaction amongst MOD organisations and between the MOD and the contractor in our proposed management model. In other cases, such as the level of personnel required in the ASM-IPT, the uncertainty surrounding workloads leads us to provide a range of possible resource requirements. These ranges are based on our interviews and consideration of the resources of similar organisations, such as those in the US nuclear submarine community.

Could the MOD's nuclear submarine-related organisations function in our proposed management model with their current level of resources? Probably—especially if the ASM-IPT grows as anticipated. But the MOD should maintain a pool of skilled and qualified personnel who can be moved amongst organisations when and as needed. As indicated by our estimates, we believe this pool of skilled nuclear submarine designers, engineers, and managers should be slightly larger than currently exists.

The MOD's Risks and Responsibilities

UK submarine acquisition has followed two very different approaches. Up through the Vanguard class, the MOD was very involved in design, acquisition, and support. With the Astute class, many of the tasks formerly conducted in-house were outsourced to the contractor (although this approach has evolved to increased ASM-IPT involvement).

Both of these approaches had benefits and costs, and both provide lessons for an analysis of the capabilities that the United Kingdom should invest in for best-practice submarine acquisition. However, we did not want to focus on fixing only the current capability gaps or problems the organisation might be facing, but instead considered it important to take a broader approach. This is based on identifying what risks the MOD cannot outsource and the responsibilities it will always have. The management model must be designed to fulfil these responsibilities and manage these risks.

The MOD will always address the following sources of risk:

- Obtaining the specified military performance
- Maintaining safety of operations
- Delivering on schedule and at cost.

The MOD is not alone in shouldering the management of these risks. The shipbuilder also faces risk if it does not efficiently deliver a cost-effective submarine; however, while the shipbuilder may go out of business, the MOD is still responsible for the defence of the nation. Again, there are risks to the company if the submarine is unsafe, but the MOD is ultimately responsible to the sailors. The MOD cannot devolve the management of these risks to contractors.

To manage these risks, the MOD must assume the following responsibilities:

- Work with industry and Customer 1 in leading submarine design, development, and downselect.

- Assess safety and technical issues in accordance with the MOD's policy that safety risks should be as low as reasonably practicable.
- Engage in monitoring the build process to ensure that ships are delivered on schedule and at projected cost.
- Develop assurance of submarine construction quality and acceptability in support of vessel acceptance.
- Ensure through-life ship safety and maintenance, and post delivery control of design intent.

Economic Theory Basis for a New Management Model

We next lay out an approach for managing the acquisition of nuclear submarines that features close engagement in the design and construction process, balancing risks and responsibilities between the MOD and the submarine industrial base. This partnership model is based on an extensive literature review and is strongly supported by interviews conducted in the United Kingdom that described existing capability gaps. The literatures we reviewed included work on organisations and on best practices in purchasing and supply chain management.

Two Approaches to Organising

Economists, organisational theorists, and sociologists have long dealt with questions of organisation, and, in this case, insights from those fields can provide context for the United Kingdom's submarine acquisition history to help point the way forward. Transaction cost economics (TCE)[1] offers a useful model to understand the United Kingdom's acquisition history. It contrasts two forms of organisation: hierarchical structures and markets. TCE can provide insight into

---

[1] Oliver E. Williamson, 'The Economics of Organization: The Transaction Cost Approach', *American Journal of Sociology*, No. 87, 1981; Oliver E. Williamson, *The Economic Institutions of Capitalism*, New York: The Free Press, 1985. These works built on the earlier work of Ronald Coase, 'The Nature of the Firm', *Economica*, No. 4, November 1937.

two contrasting acquisition approaches used by the MOD, from owning production in house (Vanguard model) to buying submarines at almost arm's length (initial Astute model). The original TCE work has been extended to incorporate intermediate forms of organisation, which is what we recommend.

TCE focuses on the trade-off between a firm's in-house production of inputs and the acquisition of those inputs from the market. Producing a good involves costs outside the directionary production costs, including management and administrative costs. Buying items on the open market involves paying not just the purchase price but also 'transaction costs' of monitoring and ensuring good performance. Furthermore, there are purported benefits to each model, including increased control in hierarchies and the benefits of market competition from outsourcing. TCE argues that firms organise as integrated hierarchies or outsource their inputs, depending on which alternative is less costly. Transaction costs are primary drivers of organisational structure, as firms form hierarchies to avoid the costs of monitoring and enforcing contracts. Transaction costs are high when there is greater frequency of exchange, quality is uncertain, and asset specificity[2] is high. Both approaches have inherent risks and costs.

Applying the Paradigm to the MOD's Experience

Using TCE as a paradigm for the MOD's experiences in submarine procurement offers a useful way of understanding the contrast between the historical procurement approach and the original Astute-class approach. This is true even though the MOD's experiences do not exactly fit the models. The MOD did not perform all the work in-house under the historic approach, since it hired design contractors and the shipbuilder.[3] And with the Astute class, the MOD has always

---

[2] Asset specificity describes investments that trading partners make that can only be used in that particular exchange relationship.

[3] It should be noted that in 1977 the UK shipbuilding industry was nationalised and the shipyard at Barrow was part of the warship sector of British shipbuilders. The industry was then privatised in the early 1980s. See Keith Hartley, *The UK Submarine Industrial Base: An*

set up mechanisms for interactions and ongoing information exchange between the government and the shipbuilder. But both the historic approach and the original Astute-class approach were close enough to the hierarchical and contract models to make the application of those models instructive.

At first glance, given the complexity of the relationship, vertical integration, as in the Vanguard model, is a reasonable approach. The experience with that approach shows that hierarchical structures have risks of their own. One big issue is that of the costs of administration and bureaucracy, which can be quite high. MOD history bears this out: It used a large bureaucracy consisting of several organisational units to manage shipbuilding. Another issue is the potential weakening of incentives for innovation. For example, in large organisations, individuals may not be motivated to innovate, because the benefits from innovation typically go to the organisation, not the individual. Some of those we interviewed did express concern about the lack of innovation in the Vanguard approach.

By the time the Astute-class contract was signed in 1997, the MOD's approach to acquisition was swinging strongly towards reliance on the contractor. As explained above, this was not a response to economic forces, but rather to political imperatives. The MOD bureaucracy had been significantly reduced, and there was a growing sense that competition could solve some of the problems inherent in the Vanguard approach. The original Astute-class model was not really an arm's-length contract, but in our interviews we heard that that market model mind-set was at least partly responsible for the structuring of the original relationship and how the early years of the programme played out. And the market mind-set is in part responsible for the decision to devolve risks of the programme to the contractor, with the problematic outcomes already noted. Also, because the situation was one of bilateral monopoly, the benefits of market competition were elusive.

Economics Perspective, York, England: Centre for Defence Economics, University of York, 2001.

An Alternative to the Hierarchical and Contract Models

Later work[4] suggests that there are alternative organisational structures with some of the features of a hierarchy and some of the features of a market. (There is disagreement as to whether this is an intermediate form of organisation or an entirely different model.) Commonly referred to as 'networks' or 'partnerships', these are cases in which the organisations have close and ongoing ties, and trust replaces the detailed monitoring typical of the pure contract form. This type of exchange allows organisations to avoid some of the problems of bureaucracy while controlling the risks of the open market. Partnerships typically feature high levels of interaction and information sharing in an environment where problems are viewed as something to be addressed by both parties and where the solutions benefit both as well. In his study of long-term business relationships, Dore referred to the 'spirit of goodwill' that formed the basis for these relationships.[5]

Why move towards a partnership approach? Podolny and Page[6] contrast formal contracting relations with a more engaged approach and suggest that in a market relationship, even if a contract is written to incorporate the possibility of future changes, some measure of trust is needed so that when the time comes, both parties will act in good faith. Given the necessity of trust, this line of reasoning suggests that trading partners should set up their initial relationships focusing on the need for that trust, and act accordingly.

There are too many examples in the literature of the benefits of partnership linkages to survey fully, but a few examples are illustrative of how close strategic relationships with high levels of information

---

[4] See, for example, Oliver E. Williamson, 'Comparative Economic Organization: The Analysis of Discrete Structural Alternatives', *Administrative Science Quarterly*, Vol. 36, No. 2, June 1991, and Walter W. Powell, 'Neither Market nor Hierarchy: Network Forms of Organization', in Barry M. Staw and L. L. Cummings, eds., *Research in Organizational Behavior*, Volume 12, Greenwich, Conn., USA: JAI Press, 1990, pp. 295–336.

[5] Ronald Dore, 'Goodwill and the Spirit of Market Capitalism', *British Journal of Sociology*, Vol. 34, 1983.

[6] Joel M. Podolny and Karen L. Page, 'Network Forms of Organization', *Annual Review of Sociology*, No. 24, 1998.

exchange, amongst other features, can lead to improved performance. Some researchers examining the biotechnology industry[7] find that close interfirm ties are a critical mechanism for learning. The reliability of information improves where ties are closer. This line of research emphasises learning as a *process*, where closely tied firms can create a community in which the development of new knowledge is enhanced. Furthermore, innovation is more likely to occur when there are strong interorganisational networks. Other researchers focus on product quality. In his research on the garment industry, Uzzi[8] finds that close ties in the form of long-term subcontracting relationships improve quality because the communication between the two parties tends to have richer information on quality-related issues. Other industries in which close ties have resulted in positive outcomes include automobile manufacturing[9] and metal machining for defence systems.[10]

Support for a Partnership Model from Best Commercial Practices

The institution of partnership arrangements is a step that has been taken in many industries, and it fits in with a growing amount of research on purchasing and supplier management. Such relationships may be viewed as analogous to the MOD's relationship with BAE Systems, its supplier of submarines.

---

[7] Walter W. Powell, Kenneth W. Koput, and Laurel Smith-Doerr, 'Interorganizational Collaboration and the Locus of Innovation: Networks of Learning in Biotechnology', *Administrative Science Quarterly*, No. 41, 1996; Walter W. Powell and Peter Brantley, 'Competitive Cooperation in Biotechnology: Learning Through Networks?' in Nitin Nohria and Robert G. Eccles, eds., *Networks and Organizations: Structure, Form, and Action*, Boston: Harvard Business School Press, 1992.

[8] Brian Uzzi, 'Networks and the Paradox of Embeddedness', *Administrative Science Quarterly*, Vol. 32, 1997.

[9] Jeffrey Dyer, *Collaborative Advantage: Winning Through Extended Enterprise-Supplier Networks*, New York: Oxford University Press, 2000.

[10] Maryellen R. Kelley and Cynthia R. Cook, 'The Institutional Context and Manufacturing Performance: The Case of the U.S. Industrial Network', National Bureau of Economic Research Working Paper 6460, March 1998.

Much literature describes the best practices that commercial firms use to manage their supplier base.[11] There are some consistent themes. Researchers in the field recommend taking a proactive approach to managing the supply base. Firms should acknowledge the strategic role that their suppliers play in areas such as quality, delivery schedule, and profitability. Work in this area views suppliers as being integral parts of the value stream for product delivery and also regards the prime contractor—and even the user—as being part of the value stream.[12]

This is not a prescription to develop partnerships with all suppliers. Not all suppliers are equally important in the delivery of a customer firm's final product, so part of the recommended strategy is to 'segment' the supply base. A number of researchers[13] recommend dividing suppliers into categories based on multiple dimensions and viewing the total amount spent on business with different suppliers in light of strategic importance, risk, or some other relevant characteristic. The firm then identifies appropriate strategies for each of the categories that results. It is more critical for companies to carefully manage suppliers that are involved in high-risk market segments and that are furnishing products on which the firm spends a lot, perhaps

---

[11] There are too many books and articles on this subject to offer a complete list here, but see, for example, Robert M. Monczka, Robert J. Trent, and Robert B. Handfield, *Purchasing and Supply Chain Management*, 2nd edition, Cincinnati, Ohio, USA: South-Western College Division, Thomson Learning, 2002; Joseph L. Cavinato and Ralph G. Kauffman, eds., *The Purchasing Handbook: A Guide for the Purchasing and Supply Professional*, 6th edition, New York: McGraw-Hill, 1999; John Gattorna, ed., *Strategic Supply Chain Alignment: Best Practices in Supply Chain Management*, Gower, England: Andersen Consulting, 1998; and Timothy M. Laseter, *Balanced Sourcing: Cooperation and Competition in Supplier Relationships*, San Francisco, Calif., USA: Jossey-Bass, 1998. See also Nancy Y. Moore, Laura H. Baldwin, Frank Camm, and Cynthia R. Cook, *Implementing Best Purchasing and Supply Management Practices: Lessons from Innovative Commercial Firms*, Santa Monica, Calif., USA: RAND Corporation, DB-334-AF, 2002, which documents specific cost saving and performance improvements that have resulted from strategic supply chain management.

[12] James P. Womack and Daniel T. Jones, *Lean Thinking: Banish Waste and Create Wealth in Your Corporation*, New York: Simon & Schuster, 1996.

[13] Two of the first were M. Bensaou, 'Portfolios of Buyer-Supplier Relationships', *Sloan Management Review*, Vol. 40, No. 4, Summer 1999, and Christopher S. Tang, 'Supplier Relationship Map', *International Journal of Logistics: Research and Applications*, Vol. 2, No. 1, 1999.

partnering with them to work on reducing cost and increasing quality. How does this apply to the MOD? A closely engaged approach is generally recommended for high-cost, high-risk items. Recognising the cost and risk inherent in submarine acquisition, the MOD should dedicate more money, time, and skilled, experienced effort to manage the procurement of submarines than it does to manage the procurement of lower-cost, lower-risk items.[14]

We see a number of benefits to a true partnership approach for the Astute-class programme. For example, increasing the flows of information will provide the MOD with a better understanding of the product it will receive. Interaction between the MOD and the contractor has the potential to help the contractor develop a product that better fits the MOD's needs. This does not mean that industry gives up design authority (the MOD may not have opinions on most of the decisions that the design authority must make). And it does not mean that the MOD makes the decision and 'takes back risk'. Rather, the relationship that would work the best is one in which the contractor has very deep insight into the MOD's needs, and this information is used to make a better product. Conversely, the MOD should have very deep insight into the design and build processes. But to gain the benefits, the MOD and the submarine contractor will have to develop a relationship based on trust and in 'the spirit of goodwill'.

Required MOD Capabilities by Acquisition Phase

We address capability requirements in two ways: by describing how MOD should manage the project over the different acquisition phases and what capabilities are needed to do this, and by suggesting enhancements to the MOD's organisation that will help it fulfil its roles. In describing the MOD-contractor partnership as it should work, we do not mean to imply that all the capabilities we identify as

---

[14] An obvious implication of this is that the Smart Acquisition approach could be tailored so that the appropriate procurement approach varies by product category.

necessary to implement that partnership are new. The MOD and the ASM-IPT currently have many in place. However, some of those capabilities could be usefully enhanced.

We have suggested that more formally embracing a partnership model will help the MOD to manage its risks and responsibilities. However, the model does not imply equal partnership in each of the acquisition phases. The MOD should manage some activities, while the shipbuilder more appropriately manages other responsibilities. Specifically, the ASM-IPT (or its successor)[15] should participate to some extent in all the phases and can serve an integrative function across the government. Table 4.1 summarises the recommended MOD role in each phase of the nuclear submarine acquisition process (phases are as described in Chapter Two). Capabilities needed are then discussed in light of these roles.

Requirements Generation

The management of requirements determination[16] in the partnership model follows the process currently in place with the increased interaction of additional submarine-related organisations within the MOD. The development of the requirements for a new class of submarines should be led by the MOD. The DEC, also known as Customer 1, generates the requirements after assessing military needs and determining the concepts of operation for the new submarine. Although acting as Customer 1, the DEC and most of his staff are Royal Navy officers, thus bringing into the requirements process the views and opinions of the fleet. During the requirements generation process, the initial requirements and concepts are generally passed to

---

[15] We use the ASM-IPT as the organisation responsible for a new submarine's design and procurement programme. For future classes of submarines—for example, the Maritime Underwater Future Capability (MUFC) or a new SSBN class—an IPT comparable to the ASM-IPT may be created.

[16] The concept and assessment stages of the CADMID cycle are comparable to the requirements and initial studies phases of our submarine life cycle.

Table 4.1
Recommended MOD Responsibilities
by Acquisition Phase

Acquisition Role	MOD Role
Requirements generation	Lead
Initial studies	Lead partner
Detailed design	Follow partner
Construction	Follow partner
Acceptance	Lead partner
Support	Lead

the DPA's Future Business Group (FBG). The FBG further develops these requirements and concepts and helps to write the URD.

Theoretically, a set of requirements that could be met by a submarine might be met by a set of other weapon systems (e.g., satellites plus aircraft plus some other systems). However, it is unlikely that a set of requirements generated by a perceived need for a new submarine would be met by anything else. Therefore, by the time a requirement for something that looks like a submarine is delivered to the FBG, it would be reasonable to get the other submarine-related organisations within the MOD involved. This would allow submarine-specific skills to be drawn on early and on an ongoing basis. For example, early involvement of an IPT managing a submarine procurement programme (i.e., the ASM-IPT[17]), the NP-IPT, and the SUB-IPT could ensure that lessons learned from ongoing construction and from the fleet are incorporated into the requirements for new classes of ships. In addition to representation from the submarine-related IPTs, the technical authorities in the STG and subject matter experts available to the MOD (e.g., at Qinetiq or the Defence Science and Technology Laboratory) should also provide inputs into the existing or future technologies that could produce

---

[17] It is possible that a submarine procurement-related IPT would not exist during the requirements phase because of a gap in submarine production. In such a case, experts from the other submarine-related IPTs might provide the basis of a new IPT formed during the initial studies phase of a new programme.

desired mission capabilities. This early involvement of submarine-related IPTs and other technical experts is a proposed change to the current model.

As the requirements determination process continues, the FBG contracts with industry groups—which may include the shipbuilder—to help define and evaluate options. Refinement of options and the requirements blends into the next phase of the process, the initial studies.

Initial Studies

The role of the MOD during the initial studies phase of a programme in the partnership model is basically the same as exists in the current model. However, there are two important changes: an increased integration of MOD resources during the early phases of the studies and an increased role for the prime contractor and/or shipbuilder during the later stages. During the course of the initial studies, the interaction between the MOD, ultimately the new IPT, and the contractor and/or shipbuilder increases.

As the studies phase begins, the FBG forms a team comprising several members of the FBG plus individuals recruited from within the MOD (or, occasionally, from outside). We recommend that one member from each submarine-related IPT be assigned to the FBG team. The team grows during the concept studies and forms the basis of the new IPT.

Concept studies are intended to generate alternative solutions to the requirements and should be managed by the MOD—primarily by the FBG. Studies can be conducted internally or by issuing contracts to specialists (which may include the shipbuilder). The FBG should lead the concept studies and have access to the capabilities needed during this phase, which include a combination of ship operators working with submarine designers and engineers with an overarching view and knowledge of design and design impact. The FBG needs enough expertise to judge the adequacy of outsourced concept studies, as well as to conduct any in-house studies that might be necessary. The skill base will also be used in the subsequent acquisition phases, although the precise set of skills required varies with the phase.

For this phase, the submarine-related organisations in the MOD currently have the skills needed to assist in concept study conduct and management and, where they do not, can draw on those of other authorities within the MOD. In fact, the FBG should initially serve as the integrator of insights from other stakeholders within the MOD, including the WSA in-service support functions, the flotilla, and the DEC. It is also at this point in the initial studies that an IPT is formed (if an appropriate one does not already exist).

Feasibility studies follow closely on the concept studies and, again, should be managed by the MOD, either by the submarine procurement–related IPT or by the FBG if the IPT is not yet formed. These studies can be conducted either internally or by support contractors, primarily the shipbuilder. At the completion of the feasibility studies, the decision can be made more comfortably as to which concept is most likely to meet the cost-capability needs of the MOD.[18]

The feasibility analysis requires a larger team than that for the concept studies, and one with more detailed submarines systems knowledge. Many of the experts used by the MOD to conduct or judge the concept studies could also be used to conduct or assess the outcomes of the feasibility studies, with the addition of other skills as needed. An important addition during the feasibility studies is representation from the DPA's Pricing and Forecasting Group (PFG) to provide cost analysis capabilities. We recommend that at least one member of the PFG be assigned to the procurement IPT when it is formed.

The MOD leads specifications development, with the submarine procurement IPT playing the lead partner role. A submarine class specification is a series of requirement statements that should be both quantifiable and verifiable.[19] The setting of the ship specification is a

---

[18] Importantly, at this point, the actual construction process is still largely an unknown risk, unless the ship is a repeat or has had heavy involvement of builders in the feasibility studies.

[19] As examples, modern nuclear submarines have been designed to be very quiet—that is, to have platform noise levels very close to ambient ocean levels (the emitted sound levels are

detailed administrative and technical process requiring essentially a system-by-system review of the new concept submarine. As noted earlier, the review and correction of prior class deficiencies are important in the submarine specification, since these specifications will form the basis for the contract to build.

The specifications development stage is where the prime contractor or shipbuilder[20] should become deeply involved and is where the partnership begins in earnest. Both parties must understand and agree on the language of the specifications. If it is possible that there will be a competition to build the eventual submarine, the MOD should work carefully with all parties to ensure that all potential partners understand the conformation of the ship.

While the specifications should be as detailed as is reasonable, it is unlikely that they would cover every contingency in future design decisions. Bounded rationality[21] on the part of the MOD and the prime contractor means that not everything can be written into the specifications, and what is written will not be perfectly understood by the shipbuilder and will thus be open to interpretation. Therefore, specifications development should mark the beginning of the development of a process for resolving issues as they arise. This will be key for the detailed design and production phases that follow.

The MOD and the shipbuilder should bring in similar suites of technical experts at this stage to ensure clear understanding of the specifications. Collocation of the MOD and the prime contractor or shipbuilder is not necessary at this phase, as long as there are sufficient meetings held to achieve mutual understanding. Once the specifications are developed and well understood, a contract can be issued. If there are multiple bidders, the selection criteria should include value (e.g., quality and reliability) and cost.

---

measured on acoustic ranges at sea). Also, individual installed components have emitted noise specifications, which are also measured to ensure that they are within specification.

[20] Other key suppliers should also be involved in the studies phase.

[21] The inability to foresee all future contingencies leads to substantial uncertainty in negotiating contracts. This uncertainty bounds the ability for the managers to act in a rational manner.

Detailed Design

In the partnership model, MOD roles during detailed design are basically as they have evolved in the past two years for the Astute programme. We suggest institutionalising the changes made by the ASM-IPT and enhancing capabilities in a few areas.

Detailed design is the first stage led by the prime contractor or shipbuilder, with the MOD—specifically the procurement IPT—as the follow partner. In this phase, the contracted specification is translated into construction and arrangement drawings for the ship. Along with setting the ship's specific features, the detailed design will drive the long-term operating and support costs of the ship.

As the follow partner, the MOD is not responsible for conducting or managing detailed design. However, there are clear benefits from engagement in the process. Bringing experts from the Naval Authorities, the SUB-IPT, the NP-IPT, and the fleet into the process means that the details of submarine design can be more closely tailored to the MOD's requirements. Even though the prime contractor may have formal design authority and hence the right to decide on trades as long as they meet the requirements, the contractor may in any given instance be indifferent to the alternatives, whereas the MOD clearly prefers one to the other. By following the partnership approach, the parties will understand that this sharing of information is not equivalent to the MOD's 'taking back risk' or taking ownership of the decision. Rather, it is a process by which the contractor acquires a richer and deeper knowledge of what its customer wants and more often makes decisions the customer would have made anyway.

An ongoing, frequent interaction on both informal and formal (i.e., design reviews) levels also means that the MOD has timely insight into design decisions and can request changes before they become firmly entrenched in the design, when modifying them would require other changes, meaning greater expense.[22]

---

[22] If the contractor has design authority, the MOD may have to pay for changes it wants.

The MOD must also be engaged in detailed design by virtue of its 'requirements ownership' responsibility. Currently, the Astute acquisition plan is structured so that the several thousand requirements laid out in the SRD are met. Grouping the requirements into areas and giving the ASM-IPT 'requirements owners' responsibility for oversight of these areas should ensure that the prime contractor delivers a product that meets these requirements. Currently, the MOD requirements owners work carefully with their counterparts at the shipbuilder on design and construction issues. Maintaining and continuing to strengthen this engagement is key to the partnership.

The skills required of the MOD during detailed design include a variety of specific design and engineering skills, as well as cost analysis expertise. These skills would allow the MOD timely insight into design so that it can understand what the contractor is doing, ensure that the design meets the requirements, and review deviations of the design from the specifications. The specific numbers of skilled people will increase from the base built up during the initial studies as the detailed design progresses. During the initial stages of detailed design, the procurement IPT should have sufficient personnel to encompass a range of naval architect, marine engineering, and specific system skills (in the next section, we will discuss the numbers of people in the IPT at various stages of the design and acquisition process). Experience is important, and the majority of these personnel should be drawn from previous submarine procurement programmes, other submarine-related IPTs, and the fleet. A contingent of five to ten personnel from the procurement IPT should be stationed at Barrow to interact with the designers.

Programme management skills are needed here as well. The procurement IPT staff needs to understand the utility and limitations of earned value management (EVM) in helping track the prime contractor's progress towards meeting its design goals. The performance-based requirements approach used in the Astute programme limits the MOD's ability to force solutions on the contractor. That is, the MOD has not specified each requirement throughout the ship to the level of detail it would have in the past, but instead has specified the performance required from the end-product submarine in critical areas. The builder is then expected to provide calculations, design, support system installation, test, and operational evidence that it has met the product requirement prior to acceptance of the vessel by the MOD. With performance-based contracting, the shipbuilder has the freedom to exploit the design trade-space from a cost-benefit perspective as long as MOD needs are met. Close ties can work to alleviate potential issues that arise from insufficient communication. Thus, in addition to the designers and engineers, the procurement IPT should have people with knowledge and expertise in programme management, contracting, and cost analysis (which would include personnel on assignment from the PFG).

Capabilities of utility to the MOD are not limited to personnel skills. An MOD investment in advanced computer modelling tools could help improve insight into the design process. Overcoming initial problems in this area, the shipbuilder has adopted computer-aided design/computer-aided manufacturing (CAD/CAM) modelling tools to help develop the detailed design and create a database, which can be of great value in construction and support. Ideally, the modelling tool will be sufficiently detailed to pick up design issues, such as overlapping parts. Programmers can also install cost-efficient design controls, such as allowing for a certain number of bends or hangers for a particular length of pipe. CAD/CAM modelling offers an opportunity for the ASM-IPT, and particularly those personnel at Abbey Wood, to stay more closely involved in the design process. There is a terminal in the ASM-IPT office, but at the time of our interviews, technical issues prevented wide use of the system. The MOD should make the necessary capability investment to be able to review the design more easily using this technology.

There is currently representation from the DEC at the ASM-IPT to provide a link to Customer 1 and to provide oversight into whether the submarine to be delivered by the ASM-IPT will meet the original requirement. This should be maintained and periodically re-examined to ensure that these objectives are being fulfilled. Currently, there are three Royal Navy officers from the DEC assigned to the ASM-IPT. We recommend that an additional Royal Navy officer be assigned to the ASM-IPT with the specific intent of stationing that officer at Barrow. (This would be preferable to moving one of the current assignees to Barrow. Below, we will argue that the current DEC contingent of three at the ASM-IPT is insufficient for acceptance purposes.)

Construction

As was the case with detailed design, the suggested MOD roles during construction are basically as they have evolved in the past two years for the Astute programme. As we will discuss in the next section, we recommend an increase in the current staff of the ASM-IPT at both Abbey Wood and Barrow.

The shipbuilder leads construction, with the MOD as the follow partner. The MOD's interest during this phase is to ensure that the submarine is built to the specifications and design and that it is finished on time and within the budget. Again, the government needs sufficient engineering skills to understand what the contractor is doing and to review questions of deviations from the design intent and as built conditions not in accordance with the detailed design (concessions). EVM is key during this phase to ascertain where the contractor stands in terms of schedule and cost on an ongoing basis.

A critical component of the production process is an integrated management control system that is managed by and available to both the shipbuilder and the procurement IPT. Building a nuclear submarine involves several major separate construction efforts, which must be synchronised and integrated: construction of the platform; of the command, control, communications, and intelligence system; of the combat system; and of the NSRP. These efforts involve the management of dozens of major milestones, hundreds of key events, and hundreds of thousands of design and construction activities. It also includes a myriad of logistics and support, construction, test, design, analysis, research and development, and engineering activities that must be managed, in addition to MOD- and contractor-managed equipment purchases.

Consider just the first stage of construction, materials acquisition. Material to support construction must be identified and sourced to vendors and placed on order with delivery dates to support the construction schedule. This involves issuing requests for proposals and letting subcontracts to suppliers. The components supplied vary in nature from the simplest lightbulbs and general service piping stock to major subsystems or such complex materials as exotic alloy piping. The government also supplies some equipment, so it needs to interface with the shipbuilder to manage delivery at the appropriate time.

The construction management control system should properly relate cost, schedule, and technical performance. It should provide technical measures to evaluate objectives and assess their execution, and it should enable the builder, together with the MOD, to monitor schedule performance with respect to all the items noted above. The combination of a good event management control system and an EVM system are important tools to which both the MOD and the builder should be committed during construction.

The decision to transfer design authority to industry knocked authority out of alignment with certain responsibilities (notably regarding risk assumption). While taking back the design authority for future classes of ships is an option for resolving this misalignment, it would be costly and may not be necessary. An alternative is for the MOD to become more involved in resolving certain kinds of issues but to leave the majority of the design authority with the prime contractor. Specifically, the MOD could provide a second check for safety-related issues, while leaving the rest of the decisionmaking with the contractor. One way of doing this is to categorise issues that arise by importance and address only the most important safety issues.

We suggest that the ASM-IPT and the prime contractor develop a hierarchy of issues and an approach for their resolution.[23] The shipbuilder could resolve the lowest-level issues by itself or with the insights of MOD personnel on site at Barrow. The shipbuilder together with the requirements owners located at Abbey Wood could address intermediate-level issues that require the inputs of the ASM-IPT. The

---

[23] The ASM-IPT does have an internal ranking of issues, but it is not clear whether the shipbuilder necessarily shares all issues with the IPT. We suggest the development of a process so that these issues are more formally brought forward.

most important questions, which predominantly relate to various elements of submarine safety, would have to be addressed by the MOD's Naval Authorities. Those would be decisions relating to submarine emergency flooding recovery systems, the pressure hull and other systems subject to test depth pressure, systems required for crew safety and ship emergency recovery, and systems required for emergency manoeuvrability. Safety-related issues need to be decided free of the commercial pressures of the shipbuilder and the cost and schedule demands confronting the ASM-IPT.[24] Requiring up-front involvement of the Naval Authorities rather than waiting for their certification near the end of the design and build process would ensure that this independent input is received at the most efficient point in the process.

MOD manning and personnel policies have contributed to effective nuclear submarine construction management. For example, increased tenure of nuclear submarine programme managers has proven beneficial in terms of profiting from teaming and from the acquired knowledge of the process and the players at every level. Keeping key management for a longer time in grade is one of the tenets of Smart Acquisition, and this is currently the policy of the MOD. The availability of active submariners for counsel and assistance during production has also proven a significant benefit.

Acceptance

We endorse the process now in place for the Astute programme with a few changes.

The procurement IPT must have the capability to gain confidence that the requirements are being met as the ship progresses through design and construction. The MOD and the shipbuilder must gain confidence that the ship is being designed in accordance with the agreed specifications derived from operational need, as described above, and that that accordance is ensured through the

---

[24] We do not mean to imply here that the prime contractor or the ASM-IPT has acted in an inappropriate manner.

actions of the design authority.[25] During construction, the process continues and additional confidence is gained through certification that the ship has been built to specification. Being engaged with the shipbuilder and developing a deep insight into the design and construction processes will help the MOD gain this confidence.

While ASM-IPT appears well organised to handle acceptance from the contractor, the DEC does not appear appropriately manned with suitably qualified and experienced personnel to handle its role in the final acceptance process. This role requires the DEC to coordinate a large number of organisations, mostly with personnel having far more experience in submarine design and construction than the fleet personnel who man the DEC on shore duty rotations. While this may work well for small, relatively simple defence systems, nuclear submarines involve a scale and complexity that far exceeds other defence systems. Similar to the concern about the lack of ASM-IPT manning over the next few years, there is also a concern about the shortage of DEC manning and experience. We recommend that the DEC augment its staff on ASM-IPT and consider hiring several experienced submarine acquisition personnel for the next few years.

To mitigate against some of the perceived weaknesses in the current acceptance system, we recommend the following actions:

1. The MOD should require IPTs to use the MCTA in the ITEA process—especially in the area of testing plans and procedure. Currently, although the MCTA is actively involved in the Astute programme, such use is only encouraged. The presence of an independent party actively involved in the early stages of acceptance planning would mute any perceptions of self-interest on the part of either the contractor or the IPT.

2. Although the number of ASM-IPT acceptance personnel at Barrow is now sufficient, we believe that the IPT would have benefited from having a larger acceptance and assurance team at Barrow earlier in the design and build process. As a guide, we suggest

---

[25] With the design authority in industry, the greater the transparency of the design authority process, the more likely the Astute will indeed be acceptable.

that the IPT have 5 to 10 personnel on hand during the design phase, 10 to 20 present during the initial stages of construction, and 25 to 35 as the programme reaches the intensive testing stages. By having more people on site earlier, the IPT leader would have a greater confidence in being able to verify that standards and requirements were being met throughout the design and build process.

3. The DEC should place a member of his staff at Barrow to provide independent verification to the CWG(Acceptance) that Astute is ready for both system and in-service acceptance. Although we have confidence that the ASM-IPT has the processes and resources in place to conduct contract acceptance, we are less confident that the DEC has the capability.

4. The MOD should ensure that BR4050, *Instructions for the Conduct of Naval Weapons Inspections and Trials*, is incorporated into future submarine contracts as the required weapons testing standard. The current Astute contract does not include this. The use of BR4050 would ensure that the contractor is using an externally recognised standard, which would reinforce confidence in the testing process.

Support

Support for in-service submarines is led by the SUB-IPT,[26] which holds in-service design authority but draws on industry for assistance.[27] The MOD needs a wide variety of engineering capabilities for this function, to analyse issues as they arise and recommend solutions, as well as the management skills to ensure that lessons learned in supporting one class of ship are transferred efficiently to supporting others and to specification development for subsequent classes.

---

[26] A planned reorganisation will bring the SUB-IPT into closer alignment with the ASM-IPT, a change discussed in the next section.

[27] In the next chapter, we will describe our rationale for transferring design authority for the Astute-class submarines that come into service to the SUB-IPT. It is our belief that, to acquire the benefits of fleetwide management efficiencies and of cross-class lessons learned, design authorities for supporting all classes of ships should be held within the same organisation.

Contracting skills to acquire appropriate inputs from industry are also needed. Also, the MOD currently uses contractors (such as DML and BES) to perform submarine servicing and refit. If a partnering approach is implemented with through-life support contractors similar to the partnership on new build programmes, similar benefits should accrue.

In the past, the shipbuilder at Barrow played little or no role in the management of the support of in-service submarines. We recommend that this change, starting with the introduction of the Astute class into service. As the design authority during design and construction, the shipbuilder has an inherent knowledge of the platform and systems that will be essential during the support phase. The shipbuilder should be part of the Submarine Support Management Group and participate in maintenance and refit decisions. We do not believe, however, that the shipbuilder should perform the maintenance and refit functions. Those are best left with DML and BES. A more effective overall integration of design skills and expertise across all submarine-related organisations will help sustain nuclear submarine design resources in an environment where new design programmes occur infrequently.[28]

In the MOD, the design authority for in-service submarines also manages the submarine configuration process.[29] Currently, the 'whole submarine design authority' is mandated to manage whole submarine characteristics, in particular hydrodynamics and weight control. Most importantly, it is the single focus for design safety and the safety certification process for in-service submarines. The organisation of the DLO's WSA design authority (in the SUB-IPT) currently supporting the Swiftsure, Trafalgar, and Vanguard classes includes design authorities to manage the whole submarine (coordination), propul-

---

[28] See the companion volume, John F. Schank, Jessie Riposo, John Birkler, and James Chiesa, *The United Kingdom's Nuclear Submarine Industrial Base, Volume 1: Sustaining Design and Production Resources*, Santa Monica, Calif., USA: RAND Corporation, MG-326/1-MOD, 2005, for elaboration on the options for sustaining scarce nuclear design and production skills.

[29] UK Ministry of Defence, Design Authority Power Section 4, *Design Authority Organisation, Responsibility and Authority*, Issue 1, 29 November 2002.

sion, ships systems and equipment, signatures, structures, power and distribution, and combat systems.

Organisational Evolution

The MOD is restructuring its submarine acquisition and life-cycle management organisations. It is creating a new organisation that brings under one umbrella the four nuclear-related IPTs: the ASM-IPT, the NP-IPT, the Nuclear Weapons IPT (NW-IPT), and the SUB-IPT. Some aspects of the current structure will apparently remain in place, particularly the relative independence of the component team leaders.

We took this change as the starting point for our recommended evolution of the submarine acquisition and support organisation. This proposal, shown in Figure 4.1, was developed principally with the current work in mind—that is, the Astute programme and support of the in-service classes of submarines—but would also be appropriate for future submarine classes. This proposal also assumes that the ASM-IPT (or any successor organisation) would manage future submarine programmes.

Roles of Various Organisations

Here, we summarise the roles of the major organisations involved in the various phases of a nuclear submarine acquisition programme. We also estimate the number and skill set of the personnel needed in each organisation. The resource levels should be viewed as initial ranges that could vary depending on the difficulty of the programme. New classes of submarines that are modest evolutions from previous classes would require fewer resources than programmes attempting revolutionary changes from previous classes.

Figure 4.1
Proposed MOD Organisation, Functions, and Interfaces

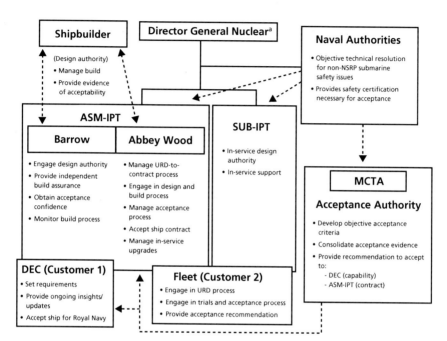

^a Organizations not shown: NP-IPT and NW-IPT.
RAND *MG326/2-4.1*

The types of skills required within the MOD are shown in Table 4.2, per expert judgement. Not all of these skills are required in each organisation. Deep technical expertise in specific areas should reside in the Naval Authorities or the science and technology laboratories. Those skills can be drawn on by the procurement IPT as outlined in the previous section.

It is important to keep a suitable number of different levels of personnel in each skill area throughout the MOD. That is, expertise should not be resident in just one or two senior individuals. There must be allowance for loss resulting from retirements and for replacement of such loss from within the MOD. For every senior-

Table 4.2
Skills Required for Nuclear Submarine Design

Naval architecture	Hydrodynamics
Marine engineering	Ship control
Mechanical engineering	Habitability
Electrical engineering	Combat systems engineering
Structural engineering	Safety and operability engineering
Shock and signatures	Testing and commissioning
Stress and dynamics	Design management
Weight engineering	Applications engineering
Metallurgy and welding engineering	Cost estimation
Radiation physics and shielding	Contracting
Systems engineering	Programme management
Nuclear propulsion	Finance
Acoustics	General management

level person, there should be two or three mid-level personnel and three or four junior personnel.

Director of Equipment Capability/Future Business Group

The DEC plays its most important role at either end of the acquisition process: requirements definition and acceptance from the ASM-IPT for the Navy. The DEC should engage with a variety of MOD submarine experts—any submarine procurement IPT that exists, the Naval Authorities, subject matter experts in science and technology laboratories, the fleet, and the WSA—as it defines the top-level requirements. Understanding and agreeing to the procurement IPT's acceptance process can give the DEC confidence to accept the submarine on behalf of the fleet.

But if the DEC is invested only at the beginning and end of the acquisition process, an opportunity for ongoing engagement of the customer in the process is missed. While the DEC has representation at the procurement IPT, and reviews requirements changes that might increase costs, augmented representation would likely provide insight into cases in which choices that affect requirements are made. We recommend that DEC representation on the IPT be increased from the current three Royal Navy officers to a contingent of four, with one of those officers stationed at Barrow. These individuals could also make a point of working with the shipbuilder in its role as design authority. The DEC's input at decision points should result in a ship that better meets the requirements.

The FBG acts as the honest broker of potential solutions to meet the requirements of the DEC. As such, it interacts with other MOD organisations with nuclear submarine expertise and with the private sector (through contracts) to identify and evaluate options to arrive at a preferred solution. We believe the current size and skill mix of the FBG to be sufficient. However, we recommend that when the FBG forms the initial team to conduct concept and feasibility studies, one or two personnel from any existing submarine procurement IPT, the NP-IPT, the SUB-IPT, and the PFG be assigned to the team. Also, skilled individuals from the Naval Authorities and the science and technology laboratories should be assigned on an as-needed basis to the team. The FBG will also need to recruit personnel from within the MOD or from outside to build the team as the studies phase progresses. This FBG-led team forms the core of the new submarine procurement IPT.

Procurement IPT (Currently ASM-IPT)

The procurement IPT manages submarine acquisition for the MOD, with the responsibility of acting as a 'supplier' delivering the submarine to the MOD customers, the DEC, and the fleet. It also manages major upgrades to existing classes of submarines. The procurement IPT has the charter as 'duty holder'[30] for submarines and as such is responsible for the through-life safety and management of the submarine.

The procurement IPT has two locations focused on submarine acquisition: one at the DPA's headquarters at Abbey Wood and the other at the shipyard, currently Barrow. The Abbey Wood location is responsible for overall programme management. Broadly, the roles there include managing the acquisition stages from the URD through the contract (with the support of other MOD organisations and con-

---

[30] UK Ministry of Defence, Joint Service Publication 430, *MOD Ship Safety Management: Issue 2, Part 2: Code of Practice*, Ship Safety Management Office, April 2003a.

tractors as appropriate), engaging in the design and build process, managing the acceptance process, accepting the ship from the contractor, and managing major in-service upgrades.

The Barrow contingent is responsible for working closely with the shipbuilder to help provide the MOD confidence regarding the company's processes and the quality of the product being delivered. Personnel should engage the design authority, work to provide independent build assurance and acceptance confidence, and monitor the build process. Engagement with the shipbuilder spans the range from technical to managerial and quality assurance. We do not propose that this engagement be directive but that it involve significant information sharing and joint problem solving. For important safety considerations, the on-site staff can collect data and share them with the procurement IPT and the MCTA.

Representatives at the yard can also play a role in protecting the programme from requirements creep or programme changes. Even the best-developed ship specifications inevitably are superseded at least in part during the construction process. While this is not necessarily the fault of the customer, it nevertheless requires discipline on the customer's behalf to avoid changing the construction specification baseline. Changes may nonetheless arise because of exogenous pressures. Because of the system density and design constraints of a submarine, the construction and cost impacts of baseline changes can be high. Each change desired by the customer must be rigorously evaluated with the builder for cost and schedule impact. It is unlikely that any requirement changes during construction will not affect other systems or equipment. Because of this, each change should be resolved between the builder and the customer as quickly as possible.

There should be 40 to 50 personnel in a new submarine procurement IPT when it is spun off from the FBG. These people should have a range of design and engineering skills as well as cost estimating expertise (from the PFG) and fleet representation (8 to 10 Royal Navy officers). Of course, there must also be management skills.

The number of personnel in the submarine procurement IPT grows as the programme progresses. When the programme reaches the detailed design phase, the IPT should have between 60 and 70 people covering the range of skills shown in Table 4.2, of whom five to ten would be stationed at Barrow.

In addition to this permanent staff, there should be one or two temporary staff from the other submarine-related IPTs and from the Naval Authorities. These personnel will not necessarily be positioned with the procurement IPT but should be available as needed to provide inputs during the design process.

The IPT continues to grow as the programme moves into the construction and acceptance phases. At its peak, the IPT should have between 110 and 120 people, with 25 to 35 at Barrow.[31] The size of the IPT, and the contingent at Barrow, is at its largest when the first of class is accepted into service. After this milestone, the IPT begins to shrink as personnel transfer to the SUB-IPT.

The ASM-IPT organisation chart as of December 2004 shows 145 personnel, distributed as shown in Table 4.3. The first data column lists the number of positions, and the second shows the number of positions occupied by Royal Navy officers.

We believe the current ASM-IPT organisation, with 74 technical personnel working directly on the Astute programme (plus the several from the PFG and DEC), when all the planned personnel slots are filled, is sufficient to be an informed and engaged customer in managing the programme at its current stage. However, ASM-IPT staffing at Barrow may be insufficient for the impending high volume of work pertaining to the acceptance process in the next few years. That should be remedied through the growth of approximately five personnel above the levels currently planned.

EVM is one particular capability that needs to be resident in a new submarine procurement IPT and strengthened in the ASM-IPT. This is used to link resource planning to schedules, incorporating

---

[31] These levels do not include personnel required to manage any major system upgrades, such as the Swiftsure- and Trafalgar-class upgrades now managed by the ASM-IPT.

Table 4.3
ASM-IPT Positions Authorised, December 2004

Section	Personnel	Of Those Personnel, Royal Navy Officers
Team leader	1	0
Executive assistant	1	0
Financial controller	8	0
Business management	23	0
Astute first buy	52[a]	8
Astute boat 4 and beyond	22	3
S&T class upgrade	32	5
From the PFG	3	0
From the DEC	3	3

[a] 34 at Barrow.
NOTE: Some of the positions in the organisation chart were not filled at the time.

technical cost and schedule requirements. Work is planned, budgeted, and scheduled in increments, with progress measured against this cost and schedule baseline. EVM offers contractors an effective cost and schedule control system and gives the government insight into how work is progressing. BAE Systems, coordinating with the ASM-IPT, is in the process of implementing an EVM system for the Astute contract. Investing in EVM oversight skills at Abbey Wood will help the government develop better insight into the programme. To accomplish this, the MOD should coordinate with BAE Systems, as well as Electric Boat, as EVM becomes more firmly established at Barrow.

Financial management of EVM can occur at Abbey Wood, and the representatives at Barrow can provide a parallel estimation of construction progress indicators. Other metrics can also be tracked to assess progress, including hours per percent progress, drawing issuance, rework, representative (cable, hanger, piping) installation rates, number of tests completed, and number of ship compartments and tanks completed and closed out. A careful distinction between leading and trailing metrics as well as between important (trend) and non-important indicators should be drawn. Critical path analysis should be conducted regularly alongside evaluation of other leading metrics.

Submarine Support IPT

The SUB-IPT is responsible for managing in-service support of the current classes of submarines. It holds design authority for these ships. (The MOD can choose to contract this authority to private industry or to maintain it internally.) We recommend that in-service design authority for the Astute class be transferred to the SUB-IPT individually when the ships enter service. That would contribute to ensuring that, for effectiveness, lessons learned from any class are transferred through the entire submarine fleet and, for efficiency, fleetwide management practices are used.

The SUB-IPT needs to engage closely with the procurement IPT to support the transfer of lessons learned to new classes of ships. Our interviews were not able to confirm that this was occurring on an ongoing basis.

Fleetwide management practices offer an example of how trust-based engagement with the shipbuilder can provide benefits to the government. Since the prime contractor holds design authority during design and build, it has the option to adopt whichever technology it chooses, as long as it meets the requirements. Under the current regime, the contractor can require the ASM-IPT to pay for changes that the government wants. Like the ASM-IPT, the SUB-IPT specifically and the WSA as a whole may prefer alternative design solutions, so that the submarine fits better into a fleetwide support management plan. For example, if all other vessels in the fleet have a particular piece of equipment, the WSA and the SUB-IPT may prefer to have that equipment as the standard on a new submarine rather than an alternative, to maintain efficiencies in such functions as logistics and repair. If the MOD and the shipbuilder have not established a strong partnership, the SUB-IPT's preference may not be accepted by the builder.

If, however, a trust-based relationship has been developed and each party has confidence that the other is trying to build the partnership, some of these concerns can be alleviated. The SUB-IPT can share information with the shipbuilder and with the ASM-IPT at the same time to transfer useful information that can be used to improve the ship.

If the SUB-IPT engages closely with the ASM-IPT and interfaces with the shipbuilder, that might also promote bringing the SUB-IPT up to speed on the CAD model used to design the Astute. This will help ease the transition of design authority to the government.

We recommend that five to ten skilled people be added to the SUB-IPT staff. This staff enhancement would help manage the Astute maintenance and refit requirements when the submarine comes into service. It can also help in our recommendation of assigning one or two people to an FBG during the studies phase of a new programme (e.g., the MUFC programme) and in assigning two or three people to a new procurement IPT.

Nuclear Propulsion IPT

The NP-IPT is located within the DPA. It primarily supports the operating fleet, and its role in Astute acquisition is limited, since the NSRP components and design services are contractor furnished. In his role as Central Plant Control Authority, the NP-IPT leader will review and approve the safety case for the Astute NSRP and will own the safety case after acceptance.

As discussed in the next chapter, having NSRP components and design services procured by the prime contractor complicates the NP-IPT's ability to prioritise allocation of technical resources within Rolls-Royce, the United Kingdom's prime source for components and technical services. Moreover, another concern is the NP-IPT's location within the DPA. Its activities appear better suited for the DLO, and the DPA's career paths emphasising general acquisition experience may be inimical to the NP-IPT's need to retain smart, experienced technical experts. It is still too early to tell whether the creation of Director General Nuclear will change this situation.

As with the SUB-IPT, we recommend that five to ten skilled personnel be added to the staff of the NP-IPT. These additional personnel could help manage the NSRP as government-furnished equipment for future submarine procurement contracts and provide the support to the FBG and the procurement IPT mentioned above.

Naval Authorities

As described earlier, the Naval Authorities reside primarily in the STG of the DPA, with two located in the DLO. Their 'authority' arises from the requirements of JSP 430, which requires them to approve safety cases prior to certain submarine operations and acceptance.

While a smart submarine designer and design authority would check with the Naval Authorities early in the design process to avoid 'surprise' problems at the end of the design and build process when safety case approvals are necessary, there is no requirement to do so. Problems found at the end of the design and build process will almost always be more expensive to fix than at the beginning of the process. Accordingly, a requirement for early involvement of the Naval Authorities in the design and build process appears judicious.

As one of the holders of detailed technical knowledge within the MOD, the staff of the Naval Authorities should be large enough to support various ship and submarine acquisition programmes. In addition to the continuation of the Astute programme, a number of new ship programmes are scheduled for the near future. These include the Future Aircraft Carrier (CVF) and the Military Afloat Reach and Sustainability (MARS) programmes. To support these various programmes and to provide expertise to the FBG and the procurement IPTs, we recommend that the staffing of the Naval Authorities be increased by five to ten skilled engineers, with approximately one-third of them knowledgeable of submarine systems.

Maritime Commissioning Trials and Assessment

As noted earlier, the MCTA provides independent assessments of acceptability to the DEC and can augment the technical expertise of the IPT. Its location within the DLO and lack of acquisition responsibilities give it independence and allow it to play an 'honest broker' role in the acceptance process.

The MCTA's greatest weakness is its lack of authority with the contractor and optional use by a naval IPT. Clarification and formalisation of its role in the acceptance process would increase its value to the MOD. To assist in submarine acceptance activities, we recommend that the staffing of the MCTA be increased by approximately five Royal Navy officers with submarine operational expertise.

The Fleet

Operator involvement is crucial for two reasons: (1) the operator is the best check on whether the submarine (or other underwater capability) under consideration will likely meet the desired need, and (2) the review of prior class documentation is unlikely to provide adequate insight into new specification needs without translation by the operator into meaningful terms. As a result, submariner input into the desired ship upgrades for both tactical and operability reasons is particularly important.

Many interviewees expressed their desire for more operator involvement. However, submariners themselves suggested that their own workforce limitations made a large permanent force of operators on site at either Abbey Wood or Barrow beyond their ability to supply. Ideally, 10 to 15 senior and experienced operators would be stationed at the submarine procurement IPT at all times, with an additional 15 to 20 at Barrow (the majority of the Barrow contingent). These could be further supported by rotations that would bring in a small number of operators for shorter periods of time (several weeks or months), which might identify others in the fleet—stakeholders on particular issues who should be engaged for an even shorter time (days). A rotation bringing in diverse submariners would increase the variety of insights and the number of people within the fleet who understood the design and operation of the new submarine sooner rather than later. Upon their retirement from active duty, these sailors could then be a source for recruitment for technical positions within the government.

Summary

In this chapter, we described the MOD's role in the acquisition process partnership and the capabilities required to support its role during each phase. To summarise, the needed capabilities include sufficient programme management, technical, and business (including independent cost estimating) resources to ensure the following:

- Sufficient design capability within MOD to partner with industry and the capability customer in leading the design, development, and downselect to production choice of future underwater capability.
- The ability to judge and decide submarine safety and related technical issues based on their merit unhampered by conflicts of interest resulting from either commercial pressure or the programme impact of decisions.
- Capability to engage in the build process to deliver ships on schedule and at project cost.
- Capability during construction to develop assurance of submarine construction quality while avoiding programme surprises, and the ability to develop objective confidence supporting product acceptance.

These responsibilities do not call for a massive increase in MOD employees. Rather, we suggest enough additional personnel within the new nuclear umbrella organisation to support increased engagement with the contractor and with other parts of the MOD, supported by sufficient resources at the DEC and within the fleet. The new partnership approach also calls for processes that will help increase and sustain trust between the government and the prime contractor. Human resources management practices that focus on hiring and promoting technical personnel (rather than generalists) will create the labour force needed to fulfil these roles into the future. However, the changes we suggest will involve a commitment from senior management within the MOD and at DPA, along with support from those who control the resources.

CHAPTER FIVE

Other Issues in Submarine Acquisition Management

Chapter Four described the roles that the MOD should fulfil during the various phases of submarine design, acquisition, and support. In this chapter, we address several issues that are important during submarine acquisition. These include the issue of design authority during the design and construction of a new class of submarines, as well as when a new submarine comes into service; the procurement of the NSRP; and the management of skilled personnel within the MOD.

The Prime Contractor as the Design Authority

While a clear and concise explication of the ship's specifications is certainly a prerequisite to common product understanding between the builder and the MOD, it does not ensure that the ship will be built as specified, or even that the specifications adequately describe the details of what goes in the ship. The design and construction processes will inevitably require adjudication of issues involving deviations from specification and departures from system design during the build process. Therefore, there exists a need to designate a point of authority for the adjudication of design issues during the submarine design and production phases.[1] This is especially important because, to a far greater extent than any other vessels, submarines operate continually in a hostile environment. Above all when submerged, continued submarine safety is dependent on sound design decisions and a body of through-life enforced specifications. The adjudication authority, called the design authority, traditionally rested with the MOD for the new design as well as for through-life support of nuclear submarines.

The design authority adjudicates questions of whether the ship's design is safe, legal, and fit for purpose. It makes sure that the ship meets the specifications (i.e., approves the final ship design) and whether an 'as-built' condition found on board the submarine under construction is acceptable (i.e., in accordance with specification). If the design does not meet the specification, the design authority can permit a 'deviation' from the design. If the as-built condition differs, the design authority can issue a 'concession' permitting this.

During design or build, the question of whether the as-built system is acceptable may frequently arise. The design authority has the final voice in this discussion. Furthermore, when facing a question of whether a 'departure from specification' is acceptable, the design authority maintains the technical competence and objectivity to address this question based on technical merit. As a result, the design authority's importance and unfettered focus are paramount in the area of submarine submerged operating safety. The design authority must, therefore, maintain technical competence and experience in the major disciplines associated with submarine design, construction, and support.

The MOD gave design authority to the prime contractor as part of the process of transferring the management of risk to the contractor. From the perspective of the customer's interests, persons exercising the design authority should be separate from those doing the design work to avoid conflicts of interest. During the design phase,

---

[1] Design issues continue throughout a project's life. In particular, this is true when technology insertion and modernisation opportunities arise. The need for a design authority therefore extends into the support phase of the submarine project as well.

ASM-IPT representatives on-site should have the capability to ensure that these two functions are truly separate and firewalled from each other. Of course, both are ultimately subordinate to top company management, whose responsibility is to make a profit for its shareholders. However, a firewalled design authority would have some opportunity to develop a corporate culture that is balanced more towards safety in the choices that must inevitably be made between product quality and cost (or schedule).

Still, the MOD must take ultimate responsibility for important safety matters. The transfer of design authority over a nuclear submarine to industry represents the creation of two distinct conflicts of interest, both of which are potential problems for submarine safety. The first conflict of interest results from the fact that industry has a primary responsibility to its shareholders to make decisions based on corporate benefit and not necessarily submarine safety.[2] Industry may be less likely than the MOD to fail-safe if it must incur higher cost when ensuring that a design or system is conservative from a safety viewpoint. When industry is the design authority, the MOD requires independent oversight and concurrence with safety-related industry design decisions. This implies, of course, that the MOD would have to be fully involved in the contractor's design process. It does not mean, however, that the MOD would usurp the contractor's design authority. The MOD's authority would vary with the importance of the issue for safety. For the most important safety-related issues, the MOD could impose its opinion on the design authority; for other issues, the MOD could work with the design authority to make its preferences known. Presumably, in a true partnership, the contractor would try to please its customer where cost-effective. If MOD's preference was more costly than the contractor's preferred alternative, the MOD could offer to pay for the change if it was sufficiently important.

The second potential conflict of interest arises because the MOD's Astute programme itself is also under both cost and schedule

---

[2] This is not, however, to suggest that there are deliberate failures on the part of industry to care about safety.

pressure. Hence within the ASM-IPT there is pressure to incur neither additional cost nor further schedule slippage. Requiring early involvement of the Naval Authorities can help ensure that design decisions regarding submarine safety receive appropriate independent oversight early in the design and build process, when design changes are far less costly.

The MOD currently does not have sufficient numbers of suitably qualified personnel to resume the role of design authority during the design and construction of a new class of submarine. As long as the conflicts of interest mentioned above can be controlled, we recommend that design authority for future submarine classes remain with the prime contractor (although, as we will discuss next, we believe design authority should revert back to the MOD when a submarine comes into service). Engagement between the procurement IPT's staff at Barrow and the contractor's design authority could provide assurance to the MOD that the design authority is exercised independently. Instead of bringing design authority back in-house, the MOD should invest in ensuring that the shipbuilder's processes are functioning correctly.

Technical Roles After Astute Goes into Service

Design authority for submarines in commission today resides in the government within the DLO's SUB-IPT. The SUB-IPT is supported by a team of contractors with experienced submarine designers and planners who maintain configuration control files and databases, plan availabilities and refit work packages, and so forth, with key technical decisions being reserved for the government. This is a natural consequence of the MOD having been the design authority for these submarines since their inception. Astute was the first submarine programme to assign this role to industry for the design and build contract.

There are advantages to retaining design authority with BAE Systems after Astute enters service,[3] namely:

- BAE Systems has an organisation fully knowledgeable about the Astute design.
- The cost of transferring the design authority function would be avoided.
- The BAE Systems organisation acting as the Astute design authority could be the cadre of submarine experienced designers necessary for a future submarine design.

Similarly, arguments can be made for transferring the design authority function to the MOD (assuming continued contractor design support) after Astute goes into service. Specifically:

- One organisation as submarine design authority could be less costly than two.
- Consistency in design practices would be ensured throughout the submarine fleet.
- Technical decisions made by the MOD would avoid commercial influence.
- Fleet operational requirements can be better accommodated by MOD personnel not concerned with corporate liability.

Several government personnel interviewed emphasised the last point concerning government personnel making decisions on commissioned ship matters. Their concern was that corporate officials are too risk averse to be making decisions concerning operating warships that must go in harm's way. If this is the case, the decision to transfer the design authority function to the MOD should be an easy one.

It should be possible to draft an invitation to tender (ITT) for a design services contract competition that could minimise data transfer costs and help preserve submarine design skills that may be needed

---

[3] The exact dates (commissioning, after the warranty period expires, etc.) are not important for purposes of this discussion.

for future submarine designs. The ITT should clearly define each party's roles, with important technical decisions reserved for the government. To avoid any commercial conflicts of interests, the ITT should require tenders to include a plan on how the bidding party would ensure that it would not benefit from any technical recommendations made to the government.

Procurement of the Nuclear Steam-Raising Plant

One area of concern regarding the effectiveness of the MOD's overall management of risk involves the NSRP. In keeping with the philosophy of making the prime contractor responsible for all aspects of the programme, the Astute prime contract required the prime contractor to subcontract with Rolls-Royce for the NSRP. As noted before, the MOD retained all nuclear risks and the PWR2 NSRP used for Astute was an existing design. In this circumstance, the prime contractor could bring no value to the NSRP design process. The prime contractor could add value if it could manage the subcontract for provision of components and information necessary for nuclear regulatory requirements better than the MOD had done previously when the NSRP was government-furnished equipment.

We collected no evidence to show whether the prime contractor managed Rolls-Royce better than the MOD had. However, even if BAE Systems were able to get priority in the allocation of Rolls-Royce's limited technical resources in support of the Astute programme, it might not be the best allocation of Rolls-Royce resources from an overall MOD standpoint. The MOD also must manage responsibilities and risks with its operating fleet, refuellings and refits at DML, and NSRP support at Faslane. Having Rolls-Royce as a subcontractor to one or more MOD prime contractors, in addition to being an MOD prime contractor, invites suboptimisation of Rolls-Royce's scarce technical resources and a diminution of the MOD's effectiveness and efficiency at managing its own risks. Considering the MOD's overall nuclear propulsion management responsibilities, it appears that having the MOD actively involved in the Rolls-Royce technical resource allocation process would be a better approach.

The management of programmes beyond Astute is outside the scope of this study. However, it appears that shifting NSRP components and design services to being government furnished on Astute would improve MOD's overall management of technical support for nuclear propulsion throughout the submarine fleet.

Concurrent Engineering

In submarine production, detailed design and construction have historically been two separate phases, and, principally for ease of exposition, we followed that traditional separation. Such a separation, however, is not currently viewed as best practice in most manufacturing processes. The current perspective is that if design and construction are integrated so that manufacturing insights and issues directly affect the design, the result should be a more easily built and hence more affordable design. This approach also calls for subcontractor input to design decisions for the same reason, especially given the significant value that comes from the supply chain typical of manufacturing.

This approach has recently been adopted for the production of the Virginia class in the United States.[4] With the transition to this type of concurrent engineering, the design process incorporates considerations related to construction and even to operation and support. Getting the manufacturing experts involved early in the process allows them to provide insight into which designs can be built less expensively. Involving support experts can bring lessons learned from other classes of boats, which can improve ease of repair and avoid the use of design alternatives that are less reliable. Operators can suggest changes to design to improve usability or access. Subcontractors can provide information on their products that can be used in developing the detailed design—or they can provide better alternatives to what

[4] General Dynamics Electric Boat, *The Virginia Class Submarine Program: A Case Study*, Groton, Conn., USA, February 2002

the designer may initially plan to use. The concurrent engineering approach has proven to have many benefits, not the least of which have been reduced requirement changes, reduced rework, and better enforcement of cost, quality, schedule, and progress standards.[5]

What would a concurrent engineering framework at the ship-builder mean for MOD's required capabilities? We do not envision it would require a significant change in numbers of people or the skill base. Rather, it would require rethinking the appropriate timing and number of review events. It would also involve getting the relevant government stakeholders and experts (including operators and those with experience in manufacturing, logistics, and other key fields) involved earlier so that they can have the appropriate input into the design process.

A Note on Communication and Trust

The suggestions for organisational change offered in the previous chapter should support improvement in the programme by increasing communication between the government and the prime contractor. The MOD should thus be more aware of problems and issues, be able to plan fleet availability workarounds sooner, and be able to provide more input. In high-risk, high-cost acquisitions such as nuclear submarines, investing in acquisition capabilities should pay off in a better-managed programme, eventually resulting in higher quality and better cost control.

However, no improvements will be possible without some attention to the nature of the linkage between the government and the contractor. The relationship between the MOD and the current contractor has not been without its difficulties; each has raised concerns about the other's behaviour. The relationship has been characterised as one that is improving, but continuing issues suggest some residual

---

[5] R. I. Winner, J. P. Pennell, H. E. Bertrand, and M. M. G. Slusarczuk, *The Role of Concurrent Engineering in Weapons System Acquisition*, Alexandria, Va., USA: Institute for Defense Analyses, R-338, December 1998.

bitterness and distrust. While this is not something that can be cured immediately, a focus on the processes by which the shipbuilder and the government engage each other can lead to a climate of increased mutual understanding and eventually trust. We suggest a task-based approach in which the MOD and the contractor would work together to identify and resolve problematic areas. For example, timely responses to design and construction issues are important. Timeliness can be improved in at least two ways. First, if the MOD would agree to provide predictable time frames, or norms, for solutions (if it is not immediately able to respond to technical questions), the shipbuilder could plan workarounds that would last as long as necessary. Second, the MOD could empower its representative at Barrow with decisionmaking authority where appropriate.

It should be noted that developing a true partnership would require willingness and participation by both the MOD and whichever contractor it chooses to work with. For example, in the current situation, BAE Systems would have to be willing to try to move beyond concerns based on any historical issues or problems and to work with the MOD on building a more constructive relationship. BAE would also need to be willing to engage the MOD in an open and trusting way, particularly in terms of sharing information on issues and problems that the company faces. Every successful resolution of specific issues for which both parties work in an open and predictable fashion will help build the foundation of a more successful tie.

Recruiting and Managing Submarine Technical Skills

Managing the DPA's nuclear qualified labour force is an important goal.[6] Plans are to maintain the skill base of this asset by consciously rotating personnel amongst the IPTs, so that they do not have to leave the nuclear submarine field and lose their qualification to be

---

[6] Sustaining the skills within the nuclear submarine industrial base for future design and production is discussed in the companion volume, Schank et al. (2005).

promoted. We suggest that other submarine-specific skills be treated in the same way, including, for example, the rotation of submarine naval architects. We also suggest that the nuclear organisation and the DPA take this plan a few steps farther. The MOD needs to aggressively manage all of its in-house technical expertise as it relates to acquisition and life-cycle management. Technically trained, skilled, and experienced people are important to the success of a shipbuilding acquisition programme.

The MOD's core technical expertise has been eroded through reduction in the ministry's workforce and the retirement of submarine specialists involved in the design and construction of earlier classes of submarines. It has also been depleted through Smart Acquisition's dual focus on job rotation for growth and on the valuation of smart generalists who can manage a contract over experts who have more technical insight into design and construction. Mid-career and senior-level experts are still available, but the formal technical career track was eliminated some time ago. There are some indications that this lack has been recognised, and attempts are being made to remedy it. However, we are not convinced that the MOD has accepted the necessity of investing in the core of technical experts required to manage the responsibilities and risks inherent in submarine acquisition.

To resolve this, we recommend that the DPA reinstate an engineering career track for submarine-related skills above and beyond nuclear-related ones. Hires can be done occasionally, with the goal of long-term career management that might include further schooling, rotations at the shipyard in Barrow, time spent temporarily at contractors, and rotations within the different nuclear-related IPTs. Currently, individuals within the DPA have strong incentives to move away from technical roles and into management jobs to be promoted. A separate technical career track in this area would provide room for growth for experts who choose to work as engineers throughout their careers. The new nuclear organisation will not be able to set policy for the DPA as a whole in this area, but we suggest that the leadership become strong advocates for this approach.

Finally, the leadership should consider collocating its subordinate nuclear IPTs within the same 'neighbourhood' at Abbey Wood. Any of the IPTs that need to be in a secure area could be located near each other, and the others could similarly be in the same open area. For example, the SUB-IPT and the ASM-IPT should be close together to increase information sharing on issues confronting new classes of submarines, upgrades of existing classes, and ongoing maintenance issues.

Managing the Transition to a New Organisation

Finally, suggesting change is a relatively simple task compared with actually implementing the changes.[7] Focusing on effective change management processes will help improvements take place. Change is difficult, and it requires significant planning and management attention. Furthermore, it is necessary to focus not just on the specific changes, but also on the environment where they take place. Senge et al.[8] describe the problem: 'The fundamental flaw in most innovators' strategies is that they focus on their innovation, on what they are trying to do—rather than on understanding how the larger culture, structures, and norms will react to their efforts'. The DPA is a complex organisation, and any change will not occur without the appropriate attention to all aspects of the change.

Change is difficult because it can be threatening: It requires organisations and their members to accept that improvements over the current way of doing things are possible. Some take this as an implicit criticism, or are not interested in learning a new approach, and thus refuse to embrace the possibility for improvement. Often

---

[7] Moore et al. (2002) describes an effective approach to change based on the best practice literature and lessons learned from large corporations.

[8] Peter Senge, Art Kleiner, Charlotte Roberts, Richard Ross, George Roth, and Bryan Smith, *The Dance of Change: The Challenge of Sustaining Momentum in Learning Organizations*, New York: Doubleday, 1999.

managers like change as a way forward, but employees find it intrusive.

There are three main phases of change. The first step is preparing for change—developing the case for change, developing the future vision, getting senior leadership support, and laying out an action plan. Executing change requires testing and validating pilot projects, deploying the change, and monitoring the outcome and refining the change. Supporting change is also necessary, requiring significant communication, the appropriate training and skill development, incentives to encourage the embracing of change, and the right level of resources.

In this report, we have laid out the case for change and have provided a future vision that resolves some of the issues we identified in the research. The next step is for MOD management to develop the way forward, recognising the difficulties that any major change involves. This by no means would be the first significant change within the MOD—we have described several others in this report. Most recently, the adoption of Smart Acquisition can provide lessons learned for what it takes to successfully implement change—and what the potential pitfalls are.

Bibliography

Bensaou, M., 'Portfolios of Buyer-Supplier Relationships', *Sloan Management Review*, Vol. 40, No. 4, Summer 1999, pp. 35–44.

Burcher, Roy, Louis J. Rydill, I. Dyer, R. Eatock Taylor, J. N. Newman, and W. G. Price, *Concepts in Submarine Design*, Cambridge Ocean Technology Series, Cambridge, England, 1995.

Cavinato, Joseph L., and Ralph G. Kauffman, eds., *The Purchasing Handbook: A Guide for the Purchasing and Supply Professional*, 6th edition, New York: McGraw-Hill, 1999.

Coase, Ronald, 'The Nature of the Firm', *Economica*, No. 4, November 1937, pp. 386–405.

Dore, Ronald, 'Goodwill and the Spirit of Market Capitalism', *British Journal of Sociology*, Vol. 34, 1983, pp. 459–482.

Dyer, Jeffrey, *Collaborative Advantage: Winning Through Extended Enterprise-Supplier Networks*, New York: Oxford University Press, 2000.

Friedman, Norman, *U.S. Submarines Since 1945: An Illustrated Design History*, Annapolis, Md., USA: Naval Institute Press, 1994.

————, *Fifty-Year War: Conflict and Strategy in the Cold War*, Annapolis, Md., USA: Naval Institute Press, 1999.

Gattorna, John, ed., *Strategic Supply Chain Alignment: Best Practices in Supply Chain Management*, Gower, England: Andersen Consulting, 1998.

General Dynamics Electric Boat, *The Virginia Class Submarine Program: A Case Study*, Groton, Conn., USA, February 2002.

Hartley, Keith, *The UK Submarine Industrial Base: An Economics Perspective*, York, England: Centre for Defence Economics, University of York,

2001. Online at www.york.ac.uk/depts/econ/research/documents/uksib. pdf (as of August 2005).

Kelley, Maryellen R., and Cynthia R. Cook, 'The Institutional Context and Manufacturing Performance: The Case of the U.S. Industrial Network', National Bureau of Economic Research Working Paper 6460, March 1998.

Kincaid, Bill, *Dinosaur in Permafrost*, Walton-on-Thames, Surrey, England, TheSAURAS Ltd., 2002.

Lacroix, F. W., Robert W. Button, John R. Wise, and Stuart E. Johnson, *A Concept of Operations for a New Deep-Diving Submarine*, Santa Monica, Calif., USA: RAND Corporation, MR-1395-NAVY, 2001.

Laseter, Timothy M., *Balanced Sourcing: Cooperation and Competition in Supplier Relationships*, San Francisco, Calif., USA: Jossey-Bass, 1998.

Macdonald, Muir, 'How Team Made an Astute Move', in UK Ministry of Defence, *Excellence in Defence Procurement: Equipping the Armed Forces*, Defence Procurement Agency, 2004, pp. 30–33.

Milgrom, Paul, and John Roberts, 'Bargaining Costs, Influence Costs, and the Organization of Economic Activity', in James E. Alt and Kenneth A. Shepsle, eds., *Perspectives on Positive Political Economy*, New York: Cambridge University Press, 1990, pp. 57–89.

Monczka, Robert M., Robert J. Trent, and Robert B. Handfield, *Purchasing and Supply Chain Management*, 2nd edition, Cincinnati, Ohio, USA: South-Western College Division, Thomas Learning, 2002.

Moore, Nancy Y., Laura H. Baldwin, Frank Camm, and Cynthia R. Cook, *Implementing Best Purchasing and Supply Management Practices: Lessons from Innovative Commercial Firms*, Santa Monica, Calif., USA: RAND Corporation, DB-334-AF, 2002.

Podolny, Joel M., and Karen L. Page, 'Network Forms of Organization', *Annual Review of Sociology*, No. 24, 1998, pp. 57–76.

Powell, Walter W., 'Neither Market nor Hierarchy: Network Forms of Organization', in Barry M. Staw and L. L. Cummings, eds., *Research in Organizational Behavior*, Volume 12, Greenwich, Conn., USA: JAI Press, 1990, pp. 295–336.

Powell, Walter W., and Peter Brantley, 'Competitive Cooperation in Biotechnology: Learning Through Networks?' in Nitin Nohria and Robert

G. Eccles, eds., *Networks and Organizations: Structure, Form, and Action*, Boston: Harvard Business School Press, 1992, pp. 366–394.

Powell, Walter W., Kenneth W. Koput, and Laurel Smith-Doerr, 'Interorganizational Collaboration and the Locus of Innovation: Networks of Learning in Biotechnology', *Administrative Science Quarterly*, No. 41, 1996, pp. 116–145.

Raman, Raj, Robert Murphy, Laurence Small, John F. Schank, John Birkler, and James Chiesa, *The United Kingdom's Nuclear Submarine Industrial Base, Volume 3: Options for Initial Fuelling*, Santa Monica, Calif., USA: RAND Corporation, MG-326/3-MOD, 2005.

Schank, John F., Jessie Riposo, John Birkler, and James Chiesa, *The United Kingdom's Nuclear Submarine Industrial Base, Volume 1: Sustaining Design and Production Resources*, Santa Monica, Calif., USA: RAND Corporation, MG-326/1-MOD, 2005.

Senge, Peter, Art Kleiner, Charlotte Roberts, Richard Ross, George Roth, and Bryan Smith, *The Dance of Change: The Challenge of Sustaining Momentum in Learning Organizations*, New York: Doubleday, 1999.

Tang, Christopher S., 'Supplier Relationship Map', *International Journal of Logistics: Research and Applications*, Vol. 2, No. 1, 1999, pp. 39–56.

UK Ministry of Defence, *Acceptance and the Achievement of New Military Capability*, no date. Online at www.ams.mod.uk/ams/content/docs/acceptan/ (as of August 2005).

UK Ministry of Defence, Design Authority Power Section 4, *Design Authority Organisation, Responsibility and Authority*, Issue 1, 29 November 2002.

———, Joint Service Publication 430, *MOD Ship Safety Management: Issue 2, Part 2: Code of Practice*, Ship Safety Management Office, April 2003a. Online at www.mod.uk/linked_files/dpa/JSP430%20Issue%202%20 Part%202%20April%202003%20Code%20of%20Pratice.pdf (as of August 2005).

———, Joint Service Publication 518, *Regulation of the Naval Nuclear Propulsion Programme*, April 2003b.

———, *The Smart Acquisition Handbook*, Edition 5, Director General Smart Acquisition Secretariat, January 2004a.

————, Joint Service Publication 430, *MOD Ship Safety Management: Issue 3, Part 1: Policy*, Ship Safety Management Office, July 2004b.

Uzzi, Brian, 'Networks and the Paradox of Embeddedness', *Administrative Science Quarterly*, Vol. 32, 1997, pp. 35–67.

Williamson, Oliver E., 'The Economics of Organization: The Transaction Cost Approach', *American Journal of Sociology*, No. 87, 1981, pp. 548–577.

————, *The Economic Institutions of Capitalism*, New York: The Free Press, 1985.

————, 'Comparative Economic Organization: The Analysis of Discrete Structural Alternatives', *Administrative Science Quarterly*, Vol. 36, No. 2, June 1991, pp. 269–296.

Winner, R. I., J. P. Pennell, H. E. Bertrand, and M. M. G. Slusarczuk, *The Role of Concurrent Engineering in Weapons System Acquisition*, Alexandria, Va., USA: Institute for Defense Analyses, R-338, December 1998.

Womack, James P., and Daniel T. Jones, *Lean Thinking: Banish Waste and Create Wealth in Your Corporation*, New York: Simon & Schuster, 1996.